# PADE APPROXIMANTS AND THEIR APPLICATIONS

# PADE APPROXIMANTS AND THEIR APPLICATIONS

Proceedings of a Conference held at the University of Kent 17–21 July, 1972

*Edited by*

P. R. GRAVES-MORRIS

*Mathematical Institute*
*University of Kent*
*Canterbury, England*

1973

ACADEMIC PRESS · London and New York

ACADEMIC PRESS INC. (LONDON) LTD.
24/28 Oval Road,
London NW1

*United States Edition published by*
ACADEMIC PRESS INC.
111 Fifth Avenue
New York, New York 10003





Library of Congress Catalog Card Number: 72-12270
ISBN: 0 12 295950 7

PRINTED IN GREAT BRITAIN BY
J. W. Arrowsmith, Ltd., Bristol

CONTRIBUTORS

F. V. Atkinson, *Mathematics Department, University of Toronto, Canada* (p. 75)

G. A. Baker, Jr., *Applied Mathematics Department, Brookhaven National Laboratory, Upton, New York 11973, U.S.A.* (pp. 83, 147)

D. Bessis, *University of Western Ontario, Canada*, and *Centre d'Etudes Nucleaires de Saclay, France* (p. 275)

J. S. R. Chisholm, *Mathematical Institute, University of Kent, Canterbury, England* (p. 11)

S. C. Chuang, *Department of Computing and Control, Imperial College, University of London, England* (p. 347)

A. K. Common, *Mathematical Institute, University of Kent, Canterbury, England* (p. 201)

H. P. Debart, *C.G.E., Marcoussis, France* (p. 351)

M. E. Fisher, *Baker Laboratory, Cornell University, Ithaca, New York 14850, U.S.A.* (p. 159)

J. Fleischer, *Los Alamos Scientific Laboratory, University of California, Los Alamos, New Mexico 87544, U.S.A.* (p. 69)

J. L. Gammel, *Los Alamos Scientific Laboratory, University of California, Los Alamos, New Mexico 87544, U.S.A.* (p. 3)

C. R. Garibotti, *Instituto di Fisica dell'Universita, Bari, Italy* (p. 253).

A. C. Genz, *Mathematical Institute, University of Kent, Canterbury, England* (p. 105)

J. Gilewicz, *Centre de Physique Theorique, Marseille, France* (p. 99)

W. B. Gragg, *Mathematics Department, University of California, San Diego, U.S.A.* (p. 117)

P. R. Graves-Morris, *Mathematical Institute, University of Kent, Canterbury, England* (p. 271)

A. J. Guttman, *School of Mathematics, University of Newcastle, Australia* (p. 163)

F. Harbus, *Massachusetts Institute of Technology, Cambridge, Massachusetts 02139, U.S.A.* (pp. 179, 183)

C. Isenberg, *Physics Laboratory, University of Kent, Canterbury, England* (p. 169)

R. C. Johnson, *Mathematics Department, Durham University, Durham, England* (p. 53)

W. B. Jones, *University of Colorado, Boulder, Colorado 80302, U.S.A.* (p. 125)

J. S. Joyce, *Wheatstone Physics Laboratory, Kings College, University of London, England* (p. 163)

A. B. Keats, *Atomic Energy Establishment, Winfrith, Dorset, England* (p. 337)

J. B. Knowles, *Atomic Energy Establishment, Winfrith, Dorset, England* (p. 337)

R. Krasnow, *Massachusetts Institute of Technology, Cambridge, Massachusetts, U.S.A.* (p. 183)

D. Lambeth, *Massachusetts Institute of Technology, Cambridge, Massachusetts, 02139, U.S.A.* (p. 183)

D. W. Leggett, *Atomic Energy Establishment, Winfrith, Dorset, England* (p. 337)

L. Lui, *Massachusetts Institute of Technology, Cambridge, Massachusetts 02139, U.S.A.* (p. 183)

C. Lopez, *Universidad Autonoma de Madrid, Spain* (p. 217)

I. M. Longman, *Department of Environmental Sciences, Tel-Aviv University, Israel* (p. 131)

D. Masson, *Department of Mathematics, University of Toronto, Ontario, Canada* (p. 41)

S. Milošević, *Institute of Physics, Belgrade, Yugoslavia* (p. 187)

J. Nuttall, *Physics Department, Texas A & M University, College Station, Texas 77843, U.S.A* and *Physics Department, University of Western Ontario, London, Ontario, Canada* (p. 29)

M. Pusterla, *Istituto de Fisica dell'Universita, Padova, Italy* (p. 299)

J. F. Rennison, *Mathematical Institute, University of Kent, Canterbury, England* (p. 271)

A. Ronveaux, *Facultés Universitaires de Namur, Namur, Belgium* (p. 135)

Y. Shamash, *Department of Computing and Control, Imperial College, University of London, England* (p. 341)

M. I. Sobhy, *Electronics Laboratory, University of Kent, Canterbury, England* (p. 321)

H. E. Stanley, *Massachusetts Institute of Technology, Cambridge, Massachusetts 02139, U.S.A.* (pp. 179, 183)

W. J. Thron, *Department of Mathematics, University of Colorado 80302, U.S.A.* (p. 23)

J. A. Tjon, *Institute for Theoretical Physics, University of Utrecht, The Netherlands* (p. 241)

G. Turchetti, *Istituto di Fisica della Universita, Bologna, Italy* (p. 313)

P. J. S. Watson, *Physics Department, Carleton University, Ottawa, Canada* (p. 93)

F. J. Yndurain, *CERN, Geneva, Switzerland* and *Universidad Autonoma de Madrid, Spain* (p. 217)

V. Zakian, *University of Manchester Institute of Science and Technology, Manchester, England* (p. 141)

## Preface

The number of scientific workers who find Padé methods useful in research seems to be growing rapidly. From the original days of the classification of some rational fraction approximants by Padé, no doubt influenced by Hermite at the Ecole Normale Supérieure, and the first theorems of de Montessus, the approximants were viewed as just another form of continued fraction, which was a well established representation. The importance of Padé Approximants as a systematic method of extracting more information from power series expansions occurring so frequently in theoretical physics was stressed by Baker and Gammel in the early 1960's. There has been a steadily growing cross-fertilisation of ideas between analysts, numerical analysts, applied mathematicians, theoretical physicists, theoretical chemists and electrical engineers since then, and this symposium is an attempt to bring together people from these diverse disciplines. The form of the colloquium was a school, primarily instructional, and a conference. The colloquium was completed by the presence of some experts who worked here for a period including both school and conference. This formal framework promoted discussions, appraisal of past work and future projects.

The topics discussed at this conference reflect current progress. The impact of Nuttall's theorem on convergence in measure and the work of the Saclay group on the perturbation series in field theory are two important developments since the publication (also by Academic Press) of "The Padé Approximant in Theoretical Physics", edited by George Baker and John Gammel. For the future, the remark in Prof. Gammel's introductory address, about the difficulties of forming Padé Approximants in several variables, may inspire the development that many of us feel is now within reach.

The conference report is laid out in five sections. The first section contains Prof. Gammel's introductory lecture and the papers on the mathematical properties of Padé Approximants. Second, there are the papers on numerical analysis and algorithms. Third, are the papers on the application of Padé Approximants to critical phenomena etc.. Fourth, come the applications in atomic, nuclear and elementary particle physics and fifth are the papers on circuit synthesis and control. Of course, the work cannot be divided into mutually exclusive sections, and many papers bear on several aspects of applied mathematics in general. The introductory lectures given at the school are being published by the Institute of Physics.

This, the conference book, is published by the camera ready process, which allows rapid publication. For reasons associated with this process, and with the requirement that the book be not too long, contributors were asked to follow some very rigid and seemingly arbitrary rules. Also, a strict length limitation was imposed. I am most grateful to the authors who complied with the instructions. The conference was directed by Prof. J. S. R. Chisholm and Dr. A. K. Common, and Dr. A. K. Common undertook the substantial task of being conference secretary. We are grateful to the Science Research Council, who sponsored the Colloquium, for funds which made the venture possible and to the Institute of Physics under whose auspices the conference was held. I wish to thank John Rennison for his encouragement during the editing, and Profs. L. Fox, J. L. Gammel, J. A. Tjon and J. L. Basdevant for academic advice. We are, as ever, grateful to our secretaries, Sallie Wilkins and Sandra Bateman for their constant and invaluable assistance.

P. R. Graves-Morris

CONTENTS

## CRITICAL PHENOMENA AND PADE APPROXIMANTS



## ATOMIC, NUCLEAR AND PARTICLE PHYSICS AND PADE APPROXIMANTS



## SIMULATION AND CONTROL

# MATHEMATICAL PROPERTIES OF PADE APPROXIMANTS

REVIEW OF TWO RECENT GENERALIZATIONS OF THE PADÉ APPROXIMANT

J. L. Gammel

(Los Alamos Scientific Laboratory, University of California

Los Alamos, New Mexico 87544, U. S. A.)

1. Quadratic Padé Approximants

The usual Padé approximant to a function $f(z)$ is the ratio of two polynomials, the numerator $N$ of degree $n$ and the denominator $D$ of degree $m$, defined by

$$Df - N = 0 \quad \text{through} \quad z^{m+n} \quad . \tag{1}$$

In other words, the approximant is the solution of a linear equation.

Recently, R. E. Shafer (1972) of the Lawrence Radiation Laboratory, Livermore, California has suggested that approximants which are the solution of higher order equations may be of value. One sees that if polynomials $P$ of degree $p$, $Q$ of degree $q$, and $R$ of degree $r$ are defined by

$$Pf^2 + Qf + R = 0 \quad \text{through} \quad z^{p+q+r+1} \quad , \tag{2}$$

and the $(p,q,r)$ approximant is taken to be an exact root of the quadratic equation, then there is a close analogy with ordinary Padé approximants. These analogies are the following.

First, the equations determining the coefficients in the polynomials $P$, $Q$, and $R$ are linear. The power series expansion of $f(z)$ is known; therefore, the power series expansion of

$f^2(z)$ is known, and P, Q, and R, and therefore their coefficients appear in Eq. (2) linearly. Because Eq. (2) is homogeneous, one of the coefficients has to be chosen. Just as the coefficient of $z^0$ in the denominator of the ordinary Padé approximants is taken to be unity, so the coefficient of $z^0$ in P may be taken to be unity. Then Eq. (2) yields $p + q + r + 2$ linear equations for the remaining coefficients of Q and R.

Second, the (N, N, N) quadratic approximant is invariant under homographic transformations

$$z = Aw/(1 + Bw) \quad . \tag{3}$$

Let $g(w) = f(Aw/(1 + Bw))$, and let $P_N[f(z)]$ be the (N, N, N) approximant to $f(z)$. Invariance means that

$$P_N[g(w)] = P_N[f(z)]\Big|_{z = \frac{Aw}{1+Bw}} \quad . \tag{4}$$

The proof of this is most easily accomplished by observing that if

$$(p_0 + p_1 z + \ldots + p_N z^N)f^2(z) + (q_0 + q_1 z + \ldots + q_N z^N)f(z)$$
$$+ (r_0 + r_1 z + \ldots + r_N z^N) = 0 \text{ through } z^{3N+1} \quad , \tag{5}$$

then substituting $z = Aw/(1 + Bw)$ and multiplying through by $(1+ Bw)^N$ yields

$$(1 + Bw)^N P\left(\frac{Aw}{1+Bw}\right)g^2(w) + (1 + Bw)^N Q\left(\frac{Aw}{1+Bw}\right)g(w)$$
$$+ (1 + Bw)^N R\left(\frac{Aw}{1+Bw}\right) = 0 \quad \text{through } w^{3N+1} \quad , \tag{6}$$

where the coefficients of $g^2$, g, and 1 are polynomials of degree N.

Since it is thought that invariance under homographic transformations greatly expands the region of convergence of ordinary [N/N] Padé approximants, so it may be expected that this invariance also greatly expands the region of convergence of (N, N, N) quadratic Padé approximants.

The third property is of importance in physics. This property is that the (N, M, N) Padé approximants to a unitary f(z) (one which satisfies $f(z)f^*(z) = 1$ for $z$ real) are unitary. The proof is pretty and proceeds as follows. By definition,

$$Pf^2 + Qf + R = 0 \quad , \tag{7}$$

and multiplying this by $f^{*2}$ and using $ff^* = 1$ ,

$$P + Qf^* + Rf^{*2} = 0 \quad . \tag{8}$$

Thus, if

$$f = \frac{-Q - \sqrt{Q^2-4PR}}{2P} \tag{9}$$

is the (N, M, N) quadratic approximant to f, then

$$f^* = \frac{-Q - \sqrt{Q^2-4PR}}{2R} \tag{10}$$

is the (N, M, N) quadratic approximant to $f^*$ . (The equality of the first and last indices is essential at this point: more generally, if Eq. (9) gives the (L, M, N) quadratic approximant to f, Eq. (10) gives the (N, M, L) approximant to $f^*$. The choice of the opposite sign for the square root is necessitated by the requirement $f$ and $f^*$ both approach unity as $z \to 0$ .)

Then clearly,

$$ff^* = 1 \text{ exactly.} \tag{11}$$

It remains to discuss the numerical power of the quadratic approximants, and this power is best illustrated by quadratic approximants to

$$\text{arc tan } z = z - \frac{z^3}{3} + \frac{z^5}{5} \cdots \quad . \tag{12}$$

To get a finite value for arc tan $z$ at $z = \infty$ using ordinary Padé approximants requires some trick such as squaring arc tan $z$, forming the Padé approximants to the resulting function of $z^2$, and taking the square root again. Then one obtains

$$\text{arc tan } z = \text{constant} - \frac{\text{constant}'}{z^2} \,, \text{ as } \; z \to \infty \quad . \tag{13}$$

This is wrong, because

$$\text{arc tan } z = \text{constant} - \frac{\text{constant}'}{z} \,, \text{ as } \; z \to \infty \quad . \tag{14}$$

Shafer's (2, 1, 2) approximant,

$$\text{arc tan } x = \frac{8z}{3 + \sqrt{25 + \frac{80z^2}{3}}} \,, \tag{15}$$

has the correct behavior at infinity, and it is much better than anything obtained from ordinary Padé approximants using so few terms.

One expects to see much research utilizing Shafer's idea.

## 2. Approximants Based on Differential Equations

One could go to cubic and quartic Padé approximants, or one might try

$$P_L \frac{df}{dz} + Q_M f + R_N = 0 \quad \text{to order} \quad L + M + N + 1 \,. \qquad (16)$$

Or one might even try second order differential equations, or in fact any sort of non-linear differential equation which one thinks appropriate.

The series related to the theory of critical phenomena should be restudied with these new ideas of Shafer's in mind. In fact, Joyce and Guttmann (1972), having come upon the idea of using second order homogeneous equations in just this way before me and independently of me, have already initiated such a program of work. I shall describe their work in a moment.

Of importance to critical phenomena is the fact that if

$$P \frac{d^n f}{dz^n} + Q \frac{d^{n-1} f}{dz^{n-1}} + \dots = 0 \,, \qquad (17)$$

then the singular points $z_c$ of f are the roots of P. If near a singular point

$$f \simeq \frac{\Gamma}{(z_c - z)^{\alpha}} \,, \qquad (18)$$

then

$$\alpha = 1 - n - \frac{Q(z_c)}{\left.\frac{dP}{dz}\right|_{z=z_c}} \,. \qquad (19)$$

There will be much in these lectures about critical points $z_c$

and critical indices $\alpha$. George Baker's trick of taking the logarithmic derivative is equivalent to the homogeneous first order differential equation method.

The expected advantage of such methods is that singularities more complicated than the one shown in Eq. (18) are in fact accommodated exactly. Consider the sum of two such singularities,

$$f = \frac{\Gamma_1}{(z_1-z)^{\alpha_1}} + \frac{\Gamma_2}{(z_2-z)^{\alpha_2}} \quad . \tag{20}$$

Differentiate once,

$$\frac{df}{dz} = \frac{\alpha_1\Gamma_1}{z_1-z}\,\frac{1}{(z_1-z)^{\alpha_1}} + \frac{\alpha_2\Gamma_2}{z_2-z}\,\frac{1}{(z_2-z)^{\alpha_2}} \quad . \tag{21}$$

Equations (20) and (21) may be solved for $(z_1-z)^{-\alpha_1}$ and $(z_2-z)^{-\alpha_2}$, (treating $\Gamma_1/(z_1-z)$ as a constant, etc.), and these results substituted into

$$\frac{d^2f}{dz^2} = \frac{\alpha_1(\alpha_1+1)\Gamma_1}{(z_1-z)^2}\,\frac{1}{(z_1-z)^{\alpha_1}} + \frac{\alpha_2(\alpha_2+1)\Gamma_2}{(z_2-z)^2}\,\frac{1}{(z_2-z)^{\alpha_2}} \quad , \tag{22}$$

and clearing away denominators one gets a second order linear differential equation with polynomial coefficients.

Combinations of powers and logarithms and powers of logarithms are accommodated: it is only a question of degree of equation and degree of polynomial coefficients required to accommodate any kind of singularities.

Joyce and Guttmann (1972) came on such methods through Joyce's observation (1972) that the Onsager solution for the free energy of an Ising model on a square lattice is related to the solution of

$$P \frac{d^2 f}{dz^2} + Q \frac{df}{dz} + Rf = 0 \quad , \qquad (23)$$

with P, Q and R of degree 9, 8 and 6 respectively.

David Gaunt, Alan Genz and I performed some calculations based on the high temperature expansion of the susceptibility of an Ising model using equations of degree as high as 3 and including the inhomogeneous term, but we found no better result for the expected critical index $\alpha=+1.75$ than has already been found by the methods to be reported by Professor Fisher. Our work was extremely superficial compared to what needs to be done, and I look forward to the output from Joyce and Guttmann's work. It is not our plan to go further with our work because of Joyce and Guttmann's priority in the field.

## References

Joyce, (1972). Private communication.

Joyce and Guttmann, (1973). "A New Method of Series Analysis", in the proceedings of this Conference.

Shafer R. E., (1972). SIAM Journal of Numerical Analysis, to be published.

CONVERGENCE PROPERTIES OF PADE APPROXIMANTS

J.S.R. Chisholm
(Mathematical Institute, University of Kent, Canterbury)

## 1. Introduction

This survey will cover some of the same ground as my final lecture on Padé Approximants given to the Canterbury Summer School; I shall also review some results which were the subjects of seminars at the Summer School. Certain aspects of convergence theory are being dealt with by other speakers: convergence and error bounds for continued fractions are being reviewed by Professor Thron and Professor Jones; Dr. Johnson is describing his own approach to the convergence problem, and convergence in certain particular applications will be dealt with by other speakers.

First, a word about notation. If the function f(z) has a power series expansion

$$\sum_{r=0}^{\infty} c_r z^r , \qquad (1.1)$$

then I denote the (N,M) Padé approximant by

$$f_{N,M}(z) \equiv \frac{\sum_{s=0}^{M} a_s z^s}{\sum_{t=0}^{N} b_t z^t} \qquad (1.2)$$

Then a row of the Padé table consists of elements $\{f_{n,M}(z)\}$, with n fixed and M = 0,1,2,.... The diagonal sequence is $\{f_{N,N}(z)\}$, while a "paradiagonal sequence" consists of the set $\{f_{N,N+j}\}$, with j fixed.

I shall discuss the convergence of three main categories of sequences:

Rows and Columns of the Padé Table
Diagonal and Paradiagonal Sequences
General Sequences $\{f_{N,M}\}$.

Several types of convergence can be discussed. These include :

(i) Convergence or uniform convergence in regions of the z-plane.

(ii) Convergence on the Riemann sphere, into which the z-plane, including the point $z = \infty$ , can be mapped one-to-one. This concept of convergence, used by Beardon (1968a) and Baker (1972) has the advantage of being invariant under the most important subgroup of the group of transformations

$$w = \frac{Az}{z+B} \quad ; \tag{1.3}$$

this is the group of transformations under which the formation of the diagonal sequence $\{f_{N,N}\}$ is invariant.

(iii) Convergence of subsequences, which can sometimes be established when a given sequence does not converge. Perron, Gammel and Wallin have given examples of row sequences and diagonal sequences which diverge at any chosen infinite sequence of points in the z-plane. Baker (1972) has shown that convergent subsequences always exist in these examples, and has discussed in detail why certain approximants have unusual properties. The reason is always the vanishing of one or more determinants of the form

$$\Delta_{m,n} \equiv \begin{vmatrix} c_m & c_{m+1} & \cdots & c_{m+n} \\ c_{m+1} & c_{m+2} & \cdots & c_{m+n+1} \\ \cdots & \cdots & \cdots & \cdots \\ c_{m+n} & c_{m+n+1} & & c_{m+2n} \end{vmatrix} . \tag{1.4}$$

When the Padé table is normal, none of these determinants is zero, and these troubles do not arise.

(iv) Convergence in measure and convergence in capacity; these are weaker forms of convergence which allow convergence except in arbitrarily small regions. Convergence in measure in a domain

D of the z-plane implies that the area in which an approximant $f_{N,M}(z)$ does not approximate f(z) within a given error can be made as small as one wishes by going far enough along the sequence. Convergence in capacity is a rather stronger result, since for any set E of points,

$$\text{meas } E \leqslant \pi \,(\text{cap } E)^2 \quad . \qquad (1.5)$$

Capacity is a linear measure, and is non-zero for a finite continuous line of points, for which the measure is zero. Smallness of cap E implies smallness of meas E.

## 2. Convergence of Rows and Columns

If $f_{N,M}$ is the (N,M) approximant to f, the reciprocal $f_{N,M}^{-1}$ is the (M,N) approximant to $f^{-1}$. So theorems on the convergence of rows of the Pade table and poles of functions can be translated directly into theorems on columns and zeros.

The first result that I shall discuss is the theorem of de Montessus de Ballore (1902) which is strongly dependent upon a result of Hadamard. I state the theorem in a slightly more general form than de Montessus did:

Let f(z) be regular in the domain $|z|<R$, except for a finite number of poles inside the domain, of total multiplicity n. Then the sequence $\{f_{n,M}(z)\}$, with n fixed, converges to f(z) for $|z|<R$, except at the poles of f(z).

de Montessus' theorem was generalised in a series of papers by Wilson (1930) to deal with functions meromorphic for $|z|<R$ , but with certain types of non-polar singularity on the circle $|z|=R$. The analysis is very heavy, but Wilson's conclusions are, stated roughly

(a) When the original power series, multiplied by the polynomial which exactly eliminates the poles in $|z|<R$, converges at a point on $|z|=R$, the approximants $\{f_{n,M}(z)\}$ converge at this point.

(b) If an approximant has more than n poles, the first extra pole "represents" the strongest non-polar singularity, the second extra pole represents the strongest remaining singularity, and so on.

In a seminar at the Summer School, Professor Saff presented a theorem which generalises and strengthens de Montessus' result. Saff's result concerns the existence and approximation of an interpolating function f(z), given the values of the function and perhaps several derivatives at a number of points in a set E in the z-plane. The set E is subject to certain restrictions related to the concept of capicity. Professor Saff has established the existence of f(z) on the set E, and the existence of approximating rational functions $f_{n,M}(z)$ with the correct total number n of poles in E, and has shown, in effect, that

$$\lim_{M\to\infty} f_{n,M} = f$$

on the set E, except at the poles. The set E can, in particular, consist of the points $|z|<R$.

Another extension of de Montessus' theorem has been given by Gragg (1972); in his theorem 8.3(b), he gives an error estimate for $f_{n,M}(z)$ for meromorphic functions f(z) with n poles inside circular disc, and at least three other poles.

In de Montessus' and Saff's theorems, the approximants have exactly the correct number of poles to match the poles of f(z). Beardon (1968b) and Baker (1972) have studied the properties of rows of the Padé table when f(z) is analytic in a circle centred at the origin, or entire. For the second row, with linear denominators, Beardon proved that a subsequence of $\{f_{1,M}(z)\}$ tended uniformly to f(z) for $|z|\leqslant \rho \leqslant R$, where R is the radius of convergence of the series (1.1). Baker's theorem 6 is a special case of this theorem, but his work elucidates the reasons why the whole row sequence may not tend to f(z), as in Perron's example.

It is harder to establish results for the general row in regions of analyticity of f(z). Beardon's theorem 4 (1968b) limits the positions of limit points of singularities of $\{f_{n,M}\}$, with n fixed, to an annulus containing the circle $|z|=R$. Baker's theorem 7 (1972) establishes convergence of the $n^{th}$ row in any closed region, for entire functions whose power series have coefficients which are "smooth" in a well-defined sense.

## 3. Diagonal and Paradiagonal sequence

These sequences are the most important. Paradiagonal sequences are related to diagonal sequences through the incorporation of a polynomial in a function f(z) and its approximants (Baker and Gammel (1961); Baker, Gammel and Wills (1961) ); so we only concern ourselves with diagonal sequences. The importance of the diagonal approximants arises from their invariance properties (Baker, Gammel and Wills (1961) ), notably invariance under transformations (1.3).

There is no known theorem of the form; "The diagonal sequence of approximants formed from (1.1) converges to f(z) with z in a domain D, if and only if ......". In the next section I shall discuss general theorems which establish weaker types of convergence. Strong convergence has only been established under certain particular conditions, and there exist counterexamples to any simple form of convergence theorem. Baker, Gammel and Wills (1961) have proposed the following "Padé conjecture", which is neither proved nor disproved :

"If f(z) is regular for $|z| \leqslant 1$, except for m poles inside the circle and at $z = 1$, with f(z) continuous as $z \to 1$ from the region $|z| \leqslant 1$, then as $N \to \infty$, a subsequence of $\{f_{N,N}(z)\}$ converges uniformly to f(z) inside $|z| \leqslant 1$ except inside small circles centred on the poles".

The special conditions assuring convergence may be

(a) conditions on the function f(z),

(b) conditions on the power series coefficients $\{c_r\}$

(c) conditions on continued fraction coefficients, say $\{a_r\}$ in the fraction

$$\frac{a_0}{z} + \frac{a_1}{z} + \frac{a_2}{z} + \ldots \qquad (3.1)$$

It is not easy to translate one type of condition (a), (b) or (c) into another type. For series of Stieltjes, the coefficients $c_r$ are of the type

$$c_r = (-1)^r \int_0^\infty u^r \, d\phi(u), \qquad (3.2)$$

where $\phi(u)$ is non-negative bounded measure function, with the series

$$\sum^\infty (c_p)^{1/2p} \qquad (3.3)$$

divergent; then the series (1.1) has the formal sum

$$f(z) = \int_0^\infty \frac{d\phi(u)}{1+uz} \qquad (3.4)$$

The function f(z) is analytic on the negative real axis $z = x \lesssim 0$; the poles and zeros of the approximants lie in this region, and the Padé approximants provide error bounds over large regions of the z-plane; Professor Jones will be discussing this in more detail. The conditions (3.2) can be related to positivity conditions on continued fraction coefficients, and can also be translated into strong sufficiency conditions on the function f(z). The convergence of continued fractions will be discussed by Professor Thron, and I shall only comment that some convergence theorems of this type relate to convergence in regions of meromorphy.

For the special class of functions

$$f(z) = a_o\, e^{\gamma z} \frac{\prod_{j=1}^{\infty} (1+\alpha_j z)}{\prod_{j=1}^{\infty} (1-\beta_j z)} \quad , \qquad (3.5)$$

with $a_o > 0$, $\gamma \geq 0$, $\alpha_j \geq 0$, $\beta_j \geq 0$ and $\sum_j (\alpha_j+\beta_j) < \infty$, with zeros at $-\alpha_j^{-1}$ and poles at $\beta_j^{-1}$, Arms and Edrei (1970) have established convergence of the diagonal sequence $\{f_{N,N}(z)\}$ .

## 4. General Convergence Theorems

I shall now discuss theorems which establish weak types of convergence for general sequences $\{f_{N,M}\}$ of Padé approximants. Baker (1970) established convergence in the union of two regions in which $\{f_{N,M}\}$ and $\{f_{N,M}^{-1}\}$, respectively, are uniformly bounded. The trouble with this theorem is that we need to know the positions of the poles and zeros of the approximants; this is a recurrent difficulty in all general convergence theorems.

There are a number of theorems dealing with the convergence of sequences $\{f_{N,M}\}$ in regions of meromorphy, except near the poles of the approximants. The first attempt to establish such a theorem was by Chisholm (1966); I showed that, if D is a closed disc of meromorphy of f(z), of radius R and centred at the origin, a sequence $\{f_{N,M}\}$ with $M,N \to \infty$ converges uniformly in a smaller disc except in arbitrarily small circular regions surrounding (i) the poles of f(z) and (ii) the limit points of poles of the sequence. It is, however, necessary to assume that the total order of poles in each approximant in the convergence region is uniformly bounded. The radius of the smaller disc depends on the positions of the poles and on the particular sequence $\{f_{N,M}\}$; if $N/M \to \infty$ and f(z) is analytic in D, the radius of the smaller disc can be slightly less than R. For the diagonal sequence $\{f_{N,N}\}$, the maximum radius is

$$r = R\ (\sqrt{2} - 1) \qquad (4.1)$$

For functions meromorphic in the z-plane, r can be taken arbitrarily large.

Beardon (1968a) improved on part of my analysis, and was able to establish similar results, but without any assumption on the boundedness of the number of poles of the approximants. A difficulty about Beardon's result is that the regions near the poles of the approximants, in which approximation breaks down, may cover the whole region in which we are seeking to establish convergence. Beardon also showed that, for functions meromorphic in $|z| < R$, this form of convergence can be established in the whole region provided (N,M) satisfies $N \geqslant k\ M$, for a sufficiently large value of $k(>1)$. For diagonal sequences, the convergence region is not the entire disc, but is limited to a smaller region.

Nuttall (1970) proved the first result which established the existence of a region in which f(z) is approximated arbitrarily closely by a member of a sequence $\{f_{N,M}(z)\}$. For functions f(z) meromorphic in the z-plane, he established convergence in measure of paradiagonal sequences $\{f_{N,N+j}\}$; in other words, he showed that the area in the z-plane in which the approximation fails can be made as small as one wishes by taking N large enough. Baker (1972) has pointed out that Nuttall's theorem, correctly stated, applies to a subsequence of a paradiagonal sequence. It seems that it would be easy to generalise Nuttall's proof to deal with more general sequences $\{f_{N,M}\}$.

Pommerenke (1972) has shown that Nuttall's theorem can be extended to establishing convergence in capacity, with the function f(z) having both poles and essential singularities, on a domain of zero capacity, in the z-plane. The inclusion of essential singularities is a notable step forward.

Zinn-Justin (1971) has provided a formulation of convergence theorems which is flexible in the sense that an arbitrary polynomial is introduced into the inequalities. The formulation, like that of Saff's theorem, depends upon contour integration. By making particular choices of the polynomial, Zinn-Justin is able to reproduce the theorems of both Nuttall and myself; I am not aware that he has reproduced the results of Beardon or Pommerenke.

It seems that further progress may be possible. Since Beardon's theorem does not specify the function f(z) in the region $|z| \geq R$, f(z) may have any kind of singularities in this region. However, Beardon did not establish convergence in measure or in capacity. Nuttall and Pommerenke, on the other hand, do not allow f(z) to have branch points. The theorem of Gammel and Nuttall, announced at a seminar in the Summer School, establishes convergence in measure for a class of quasi-analytic functions, which may have singularities dense throughout a region or along a continuous line; this theorem shows that Padé approximants can, in certain circumstances, analytically continue through a natural boundary. One would like to establish, say, convergence in capacity for a broader class of functions, so that branch points could be included. The problem of the positions of branch cuts arises, but one would hope that subsequences of the diagonal sequence would converge within circular regions containing the origin and with singularities limited to a domain of capacity zero (except, of course, at the singularities).

One can ask whether convergence in measure or in capacity is a practically useful result, since non-convergence is still possible at a countable infinity of points. In practise, points of non-convergence are very limited, except in specially constructed examples. Also, it seems that stronger forms of convergence can only be proved by imposing conditions on either the power series coefficients $\{c_r\}$ or on continued fraction coefficients. Most of the series we use in practise have some "rule of composition" of coefficients. It is worth while seeking for theorems which assume some regularity in the sequence $\{c_r\}$ - but with weaker conditions than those, for example, on series of Stieltjes - which assure stronger convergence than convergence in capacity.

## REFERENCES

Arms, R.J. and Edrei, A. (1970), in "Mathematical Essays", (Macintyre, A.J., ed.), Ohio University Bess.

Baker, G.A. Jr., (1970), Chapter 1 of "The Padé Approximant in Theoretical Physics" (Baker G.A. Jr. and Gammel J.L. eds.)

Baker, G.A. Jr., (1972), "The Existence and Convergence of Subsequences of Padé Approximants", preprint (Cornell Univ. and Brookhaven National Laboratory).

Baker, G.A. Jr., and Gammel J.L., (1961), J. Math. Analysis and Applns. 2, 21

Baker, G.A. Jr., Gammel, J.L., and Wills, J.G., (1961), J. Math. Analysis and Applns. 2, 405

Beardon, A.F., (1968a), J. Math. Analysis and Applns. 21, 344

Beardon, A.F., (1968b), J. Math. Analysis and Applns. 21, 469

Chisholm, J.S.R., (1966), J. Math. Phys. 7, 1

de Montessus de Ballore (1902), Bull. Math. Soc. de France 30, 28

Gragg, W. B., (1972), SIAM review 14, 1

Nuttall, J., (1970), J. Math. Analysis and Applns. 31, 147.

Pommerenke, C., (1972), J. Math. Analysis and Applns., to be published

Zinn-Justin, J., (1971), Proceedings of the Marseilles Symposium, Centre de Physique Theorique, C.N.R.S., Marseilles.

# RECENT APPROACHES TO CONVERGENCE THEORY OF CONTINUED FRACTIONS

W. J. Thron

(Department of Mathematics, University of Colorado, U.S.A.)

## 1. Introduction

For complex numbers $a_n, b_n, z$ define $s_n(z) = a_n/(b_n + z)$ and $S_n(z) = s_1 \circ s_2 \circ \cdots \circ s_n(z)$. Then $S_n(z)$ is a linear fractional transformation and can be written as

$$S_n(z) = \frac{A_{n-1}z + A_n}{B_{n-1}z + B_n},$$

where the $A_n$ and $B_n$ satisfy the well known second order linear difference equations.

The continued fraction algorithm is a function K which assigns to the ordered pair of sequences $\langle\{a_n\},\{b_n\}\rangle$ the sequence $\{S_n(0)\}$. $S_n(0) = A_n/B_n$ is called the nth approximant of the continued fraction. We also write $K(a_n/b_n) = \{S_n(0)\}$.

We shall emphasize here work of the past 15 years. Earlier results may be found in the books by Perron (1957) and Wall (1948) and in survey articles of the author (1961, 1964a).

## 2. The condition $s_n(V_n) \subset V_{n-1}$

The elements $a_n$ and $b_n$ of $K(a_n/b_n)$ will now be assumed to be restricted so as to satisfy the value region conditions

$$(2.1) \qquad s_n(V_n) \subset V_{n-1}$$

for a sequence of regions $\{V_n\}$. The condition (2.1) is present in most of the classical convergence results as well as in Wall's theory of positive definite continued fractions.

Let a sequence $\{V_n\}$ be given then one can determine regions $E_n$ in the product of the complex plane with itself with the property that $\langle a_n, b_n \rangle \in E_n$ implies $s_n(V_n) \subset V_{n-1}$. The region $E_n$ may be the empty set or very small. Thus a suitable choice of $\{V_n\}$ is required to obtain worthwhile results. The determination of the $E_n$ becomes somewhat easier if one requires

(2.2) $V_n \in \mathfrak{K} = [V\colon$ the boundary of the region $V$ is a circle or a straight line$]$, $n = 0,1,2,\ldots$ .

$V_n$ may contain none, part, or all of its boundary.

It is convenient to make the additional assumptions

(2.3) (a) $V_0$ is a circular disk. (b) $0 \in V_n$, $n = 1,2,3,\ldots$

Then $K_n = S_n(V_n) = S_{n-1}(s_n(V_n)) \subset S_{n-1}(V_{n-1}) \subset K_{n-1}$ so that the $K_n$ form a family of nested sets. In addition

$$S_n(0) \in K_n \subset K_{n-1} \subset \cdots \subset K_1 = S_1(V_1) \subset V_0$$

It follows that all $K_n$ are circular disks and that all approximants $S_n(0) \in V_0$. The latter information has been used in conjunction with the Stieltjes-Vitali theorem to obtain some very general convergence results. For some of these other proofs have since been found but for a number in particular those for which $V_n \notin \mathfrak{K}$, the Stieltjes-Vitali approach is still the only known method of proof. The main draw back of the method is that it yields no truncation error estimates.

Since $K_n$ is a circular disk its diameter is $2R_n$, where $R_n$ is the radius of $K_n$. Moreover, since the $K_n$ are nested, $\lim R_n = R$ always exists. If $R = 0$ we speak of the limit point case, the case $R>0$ is called the limit circle case. If $R=0$ the continued fraction must converge, if $R>0$ it may converge.

In a series of papers (Thron 1963, Hillam and Thron 1965, Jones and Thron 1968,1970) a method was developed which exploits the fact that it suffices to consider the limit circle case (convergence is assured in the limit point case) and the conditions it gives rise to together with assumptions related to the identity $s_n(\infty)=0$. With these conditions it is frequently possible to conclude that in the limit circle case $K(a_n/b_n)$ converges (this includes the possibility that the limit circle case does not occur). The advantage of the method: that it leads to a large number of convergence criteria, is in part counterbalanced by the fact that it is non constructive and thus does not lend itself to the determination of the speed of convergence. There are by now so many convergence criteria that the exploration of overlaps and containment of results has become pressing. Jones and Thron (1970) have taken some first steps in this direction.

## 3. Estimating $R_n/R_{n-1}$

In the previous section we observed that the continued fraction surely converges if $\lim R_n = 0$. This suggests trying to estimate $R_n$. For continued fractions $K(1/b_n)$ this was done by Sweezy and Thron (1967) and Field and Jones (1972) and for $K(a_n/1)$ there are results by Thron (1958,1959) and Lange (1966).

$R_n$ can be computed either by using geometric properties of linear fractional transformations or by obtaining the length of the boundary of $K_n$ by integration. In most instances it is convenient to consider $R_n/R_{n-1}$. It turns out that $R_n/R_{n-1}$ is a function of the parameters describing $V_1, \ldots V_n$, the element, $a_n$ (or $b_n$), and the quantity $h_{n-1} = S_n^{-1}(\infty) = -B_{n-1}/B_{n-2}$.

To illustrate the method we shall look at an example: Let $V_{2n} = [v: |v-a| \le \rho]$, $V_{2n-1} = [v: |v+(1+a)| \ge \rho]$ where

(3.1) $$|a| < \rho < |1+a| ,$$

Then (2.1) holds for $s_n(z) = a_n/(1+z)$ provided

(3.2) $$|a_{2n-1}^{\frac{1}{2}} \pm ia| \le \rho \quad \text{and} \quad |a_{2n}^{\frac{1}{2}} \pm i(1+a)| \ge \rho$$

Moreover
$$R_{2n}/R_{2n-1} = \frac{|a_{2n}|(\rho^2 - |1+a+h_{2n-1}|^2)}{|(1+a)h_{2n-1} - a_{2n}|^2 - \rho^2 |h_{2n-1}|^2}$$

By showing that $|h_{2n-1} + (1+ak_n)| \le \rho k_n$, $k_n = (n-1)/(n-1+b)$, $b = b(a,\rho)$, Lange (1966) was able to prove that $R_n < n^{-d}$, $0 < d = d(a,\rho)$.

By restricting $a_n$, for example by replacing $\rho$ by $\rho - \varepsilon$ and $\rho + \varepsilon$ in (3.2), respectively, much smaller values for $R_n$ can be obtained. This is an area that requires further attention. Here it must suffice to observe that some continued fractions with complex elements converge very slowly, while others converge quite fast ($R_n < cg^n$, $0 < g < 1$ has been obtained in many cases).

Finally, note that (3.1) insures among others that $0 \in V_{2n-1}$, $0 \in V_{2n}$. If (3.1) is replaced by $|a| < \rho$, $|1+a| < \rho$ then $0 \in V_{2n}$, $0 \notin V_{2n-1}$. In this case the sequence $\{S_{2n}(0)\}$ still converges (Thron 1964b). Earlier results of this nature were given by Perron (1957b) and Thron (1959).

References

Field, D. and Jones W. B. (1972). Numer. Math.

Hillam, K. L. and Thron W. J. (1965). Proc. Amer. Math. Soc. 16, pp 1256 - 1262.

Jones, W. B. and Snell R. I. (1969). SIAM J. Numer. Anal. 6, pp 210 - 221.

Jones, W. B. and Thron W. J. (1968). Canad. J. Math. 20, pp 1037 - 1055.

______. (1970). Trans. Amer. Math. Soc. 150, pp 93 - 119.

Lange, L. J. (1966). Illinois J. Math. 10, pp 97 - 108.

Perron, O. (1957a). "Die Lehre von den Kettenbrüchen", Teubner, Stuttgart.

______. (1957b). Bayer. Akad. Wiss. Math. Nat. Kl. Sb. No.1.

Sweezy, W. B. and Thron W. J. (1967). SIAM J. Numer. Anal. 4, pp 254 - 270.

Thron, W. J. (1958). Math. Zeitschr. 69, pp 173 - 182.

______. (1959). Math. Zeitschr. 70, pp 310 - 344.

______. (1961). Amer. Math. Monthly 68, pp 734 - 750.

______. (1963). J. Ind. Math. Soc. 27, pp 103 - 127.

______. (1964a). Math. Student 32, pp 61 - 73.

______. (1964b). Math. Zeitschr. 85, pp 268 - 273.

Wall, H. S. (1948). "Analytic Theory of Continued Fractions" Van Nostrand, Princeton.

# VARIATIONAL PRINCIPLES AND PADE APPROXIMANTS

by

J. NUTTALL†

(Physics Department, Texas A & M University, College Station, Texas 77843, U.S.A., and Physics Department, University of Western Ontario, London, Ontario, Canada*)

## 1. Introduction

Variational principles have played an important role in the development and formulation of the laws of modern physics. They have also been of immense value as a means of deriving approximate methods for the numerical solution the differential equations that appear in physics and other branches of Science. Particularly in the case of minimum principles, these methods may be placed on a rigorous mathematical foundation, and they are often very successful in practice.

Padé approximants also show promise of being helpful in solving a number of difficult computational problems, and recently it has become

† Supported in part by the U.S. Air Force Office of Scientific Research under Grant No. 71-1979A

* Present address

clear that there is a connection between P.A.'s and certain approximations derived from variational methods. The purpose of this review is to discuss this connection with the hope that it may lead to progress in understanding the properties of P.A.'s and make their application more successful.

## 2. Variational Principles

Let us review several important types of variational principle and, to fix our ideas, let us consider the case of quantum mechanics, where we have a Hamiltonian operator H (self-adjoint) acting in some Hilbert space. Suppose that H can be written as $H = H_o + \lambda V$, where $\lambda$ is a parameter.

### 2a. Rayleigh-Ritz Method

This method gives a means of calculating a discrete eigenvalue E of H , eigenfunction $\psi$, so that

$$H\psi = E\psi \tag{1}$$

The principle states that the quantity [E], given by

$$[E] = (\psi_t , H\psi_t) / (\psi_t , \psi_t) , \tag{2}$$

is stationary for small variations of the trial

function $\psi_t$ about $\psi$, and that when $\psi_t = \psi$, $[E] = E$. Thus, if $\psi_t = \psi + \delta\psi$, we have

$$[E] = E + 0\ (\delta\psi^2) \quad , \tag{3}$$

so that if $\psi_t$ is accurate to first order , [E] gives an estimate of E accurate to second order. This type of behaviour is common to all variational principles and is an important factor in their success.

The principle is used to provide an estimate of the eigenvalues of H by restricting $\psi_t$ to lie within a subset of the Hilbert space and finding the stationary values of [E] under this restriction. Often the subset is chosen to be a finite dimensional linear subspace spanned by the functions $u_n$ , $n = 1,\ldots,N$, with $(u_n, u_m) = \delta_{nm}$, in which case the problem reduces to finding the eigenvalues of the matrix $H_{nm} = (u_n, Hu_m)$. For the lowest eigenvalue, the Rayleigh-Ritz method is a minimum principle. A good deal is known about the convergence of approximations derived from minimum principles (Mikhlin, 1965, 1971).

A principle similar in structure to the Rayleigh-Ritz principle may be obtained for the value of $\lambda$ required to give an eigenstate of H at energy E, chosen to be less than the lowest eigenvalue of $H_0$ (Alabiso et al., 1970, 1971). It is easy to see that, if

$$\phi = (H_0 - E)^{-\frac{1}{2}} V\psi$$

then

$$(H_0 - E)^{-\frac{1}{2}} V(H_0 - E)^{-\frac{1}{2}} \phi = \frac{-1}{\lambda} \phi \tag{4}$$

which gives rise to a principle of structure similar to (2) for $-1/\lambda$, with H replaced by $(H_0 - E)^{-\frac{1}{2}} V(H_0 - E)^{-\frac{1}{2}}$.

The Rayleigh-Ritz method has been used with great success in atomic, molecular and nuclear physics and in many other areas of applied mathematics. For instance the binding energy of the Helium atom was calculated to 1 part in $10^{10}$ over 10 years ago (Pekeris, 1959). The binding energy of the triton for "realistic" nuclear potentials has been calculated reasonably accurately (Delves and Phillips, 1969).

2b. Inhomogeneous Rayleigh-Ritz Method

Suppose we wish to calculate $I = (f, (E - H)^{-1}g)$ with E complex and f, g two functions in the Hilbert space. A stationary expression for I is

$$[I] = (f, \chi_t) + (\chi_t', g) - (\chi_t', (E - H)\chi_t) \tag{5}$$

The exact values of $\chi_t$ , $\chi_t'$ are

$$\chi_t = (E - H)^{-1} g$$
$$\chi_t' = (E^* - H)^{-1} f^* \tag{6}$$

It may be shown that this method is equivalent to the Kohn-Hulthén method if E approaches a real value and f and g are chosen appropriately.

The Kohn method has been used in accurate calculations of the scattering of electrons from hydrogen atoms (Schwartz, 1961). A peculiarity of the method should be noted. As in the Rayleigh-Ritz method, $\chi_t$ and $\chi_t'$ are expanded as linear combinations of some basis functions $u_n$ and the stationary value of [I] may be written in terms of the inverse of a matrix. For certain values of E, this inverse may not exist and our estimate of I is infinite. The convergence of approximations derived from stationary variational principles does not appear to have received much attention from mathematicians, but special cases have been studied. It has been shown in certain models that the Kohn method converges in measure, so that for a sufficiently high order of approximation, the estimate is good for almost all values of E in a given range (Nuttall, 1969; Westwater, 1972).

2c. Schwinger Method

The Schwinger method calculates the amplitude T given by

$$T = (f, [V + \lambda V(E - H)^{-1} V] g) \tag{7}$$

where again E is complex. A stationary expression for T is

$$\begin{aligned}[T] = {} & (\psi_t', Vg) + (f, V\psi_t) \\ & - (\psi_t', [V - \lambda V(E - H_0)^{-1} V] \psi_t)\end{aligned} \tag{8}$$

An example of the use of a modified version of the Schwinger method is the solution of the Bethe-Salpeter equation for the exchange of a scalar meson by Schwartz and Zemach (1966).

## 3. Padé Approximants and Variational Methods

A basic formula (Baker, 1970), for the [N,M] P.A. to a series

$$F(\lambda) = \sum_{n=0}^{\infty} a_n \lambda^n \tag{9}$$

is

$$[N,M] = \sum_{j=0}^{M-N} a_j\lambda^j + \underline{w}^T W^{-1}\underline{w}\lambda^{M-N+1} \tag{10}$$

where it is assumed that $a_j = 0$ if $j < 0$. Here the $N \times N$ matrix $W$ is given by $W_{ij} = a_{M-N+i+j-1} - \lambda a_{M-N+i+j}$, $i,j = 1,\ldots,N$ and the column $\underline{w}$ by $w_i = a_{M-N+i}$, $i = 1,\ldots,N$.

We have shown that (10) also holds, in the special case $M - N - 1$, if $a_j$ is a square matrix and the P.A. is defined as $F = ND^{-1}$. Quite possibly, the generalization of (10) to matrix P.A.'s will hold for any M, N, but this has not been proved.

Let us illustrate the connection between P.A.'s and approximations derived from variational principles by considering the Schwinger method. Suppose in (8) that we restrict $\psi_t$ and $\psi_t'$ to N dimensional linear subspaces, so that

$$\psi_t = \sum_{j=1}^{N} c_j[(E - H_o)^{-1}V]^{j-1} g$$

$$\psi_t' = \sum_{j=1}^{N} c_j'^*[(E^* - H_o)^{-1}V]^{j-1} f^* , \tag{11}$$

where $c_j$, $c'_j$ are to be varied. Then [T] becomes

$$[T] = \underline{c}'^T \underline{w} + \underline{w}^T\underline{c} - \underline{c}'^T W \underline{c} \tag{12}$$

where W,$\underline{w}$ are as defined above for M = N - 1, with

$$a_j = (f, V[(E - H_o)^{-1}V]^{j-1}g) \tag{13}$$

The stationary value of [T] is given by $\underline{w}^T W^{-1} \underline{w}$, which is seen from (10) to be the [N,N - 1] P.A. to the expansion of [T] in powers of $\lambda$. The ansatz (11) was first considered by Cini and Fubini (1953, 1954).

Other P.A.'s to [T] may be obtained in a similar manner from the Schwinger and Kohn principles. The details have been given by Nuttall (1970), Garibotti (1972) and Bessis and Pusterla (1968).

The above argument may be extended to matrix P.A.'s. Suppose that there are m different functions of f, called $f^{(i)}$ say, and similarly m $g^{(i)}$. Then $T^{(ij)}$ is an m x m matrix with a power series expansion in $\lambda$ whose coefficients are the m x m matrices $a_k^{(ij)}$,

$$a_k^{(ij)} = (f^{(i)}, V[(E - H_o)^{-1}V]^{k-1} g^{(j)}) \tag{14}$$

It is not difficult to see from (10) that, if $\psi_t$

is chosen to be

$$\psi_t = \sum_{j=1}^{N} \sum_{i=1}^{M} c_j^{(i)} [(E - H_o)^{-1}V]^{j-1} g^{(i)} \quad (15)$$

with $c_j^{(i)}$ as variables, and similarly for $\psi_t'$, then the stationary value of [T] is the [N, N - 1] matrix P.A. to T.

Another connection between P.A.'s and variational principles occurs when we consider the values of $\lambda$ at which H has a discrete eigenvalue. These correspond to values of $\lambda$ for which T has a pole. The approximation to T derived from the [N, N] P.A. to T (with f = g) will have a pole when det $(M_{ij}) = 0$, where

$$M_{ij} = a_{i+j} - \lambda a_{i+j+1} \quad , \; i,j = 1,\ldots,N \quad ,$$

If the trial function

$$\phi = (H_o - E)^{\frac{1}{2}} V \sum_{i=1}^{N} c_i [(H_o - E)^{-1}V]^{i-1} g \quad (16)$$

is used in the Rayleigh-Ritz principle for (4), the same equation will result. Generalizations and more details will be found in the work of Alabiso et al., (1970, 1971).

An earlier relation between P.A.'s and variational methods for eigenvalues was given by

Young et al., (1957). They considered P.A.'s to the Brillouin-Wigner pertubation expansion for the eigenvalue. Suppose that $\lambda_0$ is a normalized eigenfunction of $H_0$, eigenvalue $E_0$, and that $\psi = \psi_0 + \phi$ with $(\phi, \psi_0) = 0$. Then it may be shown that for any trial function $\psi_t = \psi_0 + \phi_t$ used in the Rayleigh-Ritz principle (2), the estimate of E, say $E_t$, obtained satisfies the equation

$$E_t = E_0 + \lambda(\psi_0, V\psi_0) + \lambda(\psi_0, V\phi_t) + \lambda(\phi_t, V\psi_0) - (\phi_t, [E_t - H]\phi_t) \qquad (17)$$

If $\phi_t$ is chosen to have the form

$$\phi_t = \sum_{j=1}^{N} c_j \quad [Q(E_t - H_0^Q)^- V]^j \psi_0 \qquad (18)$$

where Q projects onto the subspace orthogonal to $\psi_0$, and the $c_j$ are varied to make the right-hand side of (17) stationary, it is found from (10) that $E_t$ is given by the [N,N + 1] P.A. to the Brillouin-Wigner expansion for the eigenvalue.

## 4. Discussion

One of the main uses of the relation between P.A.'s and variational methods is to help in providing a sounder basis for the calculations in quantum field theory described elsewhere in this volume. A problem here is the question of how renormalization fits into the variational method, for it is the renormalized pertubation expansion that is summed. Another rather speculative question is whether P.A.'s can be used to sum a series for the error in an estimate of a quantity derived from a variational approximation.

## References

Alabiso, C., Butera, P. and Prosperi, G.M. (1970). Nuovo Cimento Letters 3 , 831.

Alabiso, C., Butera, P. and Prosperi, G.M. (1971). Nuclear Phys. B31 , 141.

Baker, G.A. (1970). The Padé Approximant in Theoretical Physics, Academic Press, New York, p.1, edited by G.A. Baker and J.L. Gammel.

Bessis, D. and Pusterla, M. (1968). Nuovo Cimento, 54A , 243.

Cini, M. and Fubini, S. (1953). Nuovo Cimento, 10 , 1695.

Cini, M. and Fubini, S. (1954). Nuovo Cimento, 11 , 142.

Delves, L.M. and Phillips, A.C. (1969). Rev. Mod. Phys. 41 , 497.

Garibotti, C.R. (1972). Ann. Phys. (N.Y.), to be published

Mikhlin, S.G. (1965). Problem of the Minimum of a Quadratic Functional, Holden-Day Inc., San Francisco.

Mikhlin, S.G. (1971). Numerical Performance of Variational Methods, Walters-Noordhoff, Groningen.

Nuttall, J. (1969). Ann. Phys. (N.Y.) 52 , 428.

Nuttall, J. (1970). The Padé Approximant in Theoretical Physics, Academic Press, New York, p.219, edited by G.A. Baker and J.L. Gammel.

Pekeris, C.L. (1959). Phys. Rev. 115 , 1216.

Schwartz, C. (1961). Phys. Rev. 124 , 1468.

Schwartz, C. and Zemach, C. (1966). Phys. Rev. 141 , 1454.

Westwater, M.J. (1972). Preprint, University of Washington.

Young, R.C. Biedenharn, L.C. and Feenberg, E. (1957). Phys. Rev. 106 , 1151.

PADE APPROXIMANTS AND HILBERT SPACES

D. Masson

(Department of Mathematics, University of Toronto, Canada)

## 1. Introduction

There exists a close connection between the theory of the classical moment problem and Padé approximants or continued fractions on one hand and the theory of operators in a Hilbert space on the other hand. The connection was realised and developed during the 1920's (see the book by M. Stone, 1932 and the more recent survey by N. I. Akhiezer, 1965) and there has been renewed interest over the last several years in connection with applications to problems in quantum theory.

The applications have both a foundational aspect (one can prove that certain operators of quantum theory are self-adjoint) and a numerical aspect (one obtains approximation methods for the solving of eigenvalue problems and scattering problems). Before reviewing some of the mathematical background for these applications, it will be necessary to recall some of the basic facts concerning symmetric operators.

Let $H$ be a Hilbert space with inner product (f,g) and norm $||f|| = (f,f)^{\frac{1}{2}}$. A linear operator A is said to be symmetric if

(i) $D(A)$, the domain of A is dense in $H$ (given $g \in H$, $\varepsilon > 0$, one can find $f \in D(A)$ such that $||g-f|| < \varepsilon$) and

(ii) $(g,Af) = (Ag,f)$ for all $f,g \in D(A)$.

The adjoint $A^+$ of A is defined by the equation $(g,Af) = (A^+g,f)$ for fixed $g \in \mathcal{D}(A^+)$ and all $f \in \mathcal{D}(A)$. One necessarily has $\mathcal{D}(A^+) \supseteq \mathcal{D}(A)$ and $A^+g = Ag$ for $g \in \mathcal{D}(A)$. That is to say $A^+$ extends A ($A^+ \supseteq A$). It follows that for a general symmetric A one has $A \subseteq A^{++} \subseteq A^+ = A^{+++}$ where $A^{++} = \bar{A}$ is the minimal, closed, symmetric extension of A. One calls A self-adjoint (s.a.) if $A = A^+$ and essentially self-adjoint (e.s.a.) if $A^{++} = A^+$. In the latter case $\bar{A}$ is s.a.. For A to be e.s.a. (s.a.) it is necessary and sufficient for zI-A to have a densely (everywhere) defined inverse for all z with Im $z \neq 0$. The importance of the self-adjointness property lies in the fact that one then has a spectral decomposition given by $A = \int_{-\infty}^{\infty} \lambda \, dE(\lambda)$ with $E(\lambda)$ an increasing family of orthogonal projection operators. A complete knowledge of A is then obtained once its spectral family $E(\lambda)$ is known and one has the representation $f(A) = \int_{-\infty}^{\infty} f(\lambda) \, dE(\lambda)$ for functions of the self-adjoint operator A. Thus the two basic problems one must solve when confronted with a symmetric operator A are

(i) Is A s.a. or e.s.a.?

(ii) If so, can one determine its spectral family $E(\lambda)$?

In Section 2 we consider the special case when A is a Jacobi matrix and indicate the answers to both these questions. In Section 3 we show how one can generate Jacobi matrices and approximants for an arbitrary symmetric operator. The basic idea is to associate with the pair A,f, where A is a symmetric operator and f is a vector, the sequence $\{C_n = (f,A^n f)\}$ and the series $P(z) = \sum_{n=0}^{\infty} C_n Z^n$. The sequence defines a Hamburger moment problem and the series is associated formally with the power series expansion of the resolvent $(I-ZA)^{-1}$ of the operator A. If one can show that the moment problem is determinate then one obtains

(i) information concerning the self-adjointness of A and

(ii) a converging sequence of approximate operators $A_N$ and corresponding resolvents $(1-ZA_N)^{-1}$.

The resolvents are associated with the [N/N+1] Padé approximant to the power series P(Z).

## 2. Jacobi Matrices

Consider the Hilbert space

$$\ell^2 = \{x = (x_o, x_1, \ldots) \mid \sum_{n=o}^{\infty} |x_n|^2 < \infty\} ,$$

the space of square summable complex sequences with the inner product $(x,y) = \sum_{n=o}^{\infty} \bar{x}_n y_n$. An infinite tridiagonal matrix

$$A = \begin{pmatrix} a_o & b_o & 0 & \cdots & \\ b_o & a_1 & b_1 & 0 & \cdots \\ 0 & b_1 & \cdots & & \\ \vdots & 0 & & & \\ \vdots & \vdots & & & \end{pmatrix} \qquad (1)$$

with $a_n$ real and $b_n > 0$, n=0,1,... is called a Jacobi or J-matrix. Its action in the space $\ell^2$ is defined in standard fashion by

$$(Ax)_n = b_{n-1}x_{n-1} + a_n x_n + b_n x_{n+1} , \qquad (2)$$

with a domain $D(A) = \{x \mid x_n = 0$ for all n sufficiently large$\}$, the set of vectors with a finite number of non-zero components. The space $\ell^2$ is spanned by the orthonormal basis $\{e_n\}$ with

$$e_n = (0,0,\ldots,1,0,\ldots)$$

having a one in the nth component. One has

$$e_n = Q_n(A)e_o \ , \quad n=0,1,\ldots \tag{3}$$

where $Q_n$ is a polynomial of degree n and satisfies the recursion relation

$$AQ_n(A) = b_{n-1}Q_{n-1}(A) + a_nQ_n(A) + a_nQ_n(A) + b_nQ_{n+1}(A), \tag{4}$$
$$n=0,1,\ldots$$

with the initial condition $Q_{-1} = 0$, $Q_o = 1$.

Associated with the J matrix one has the sequence

$$C_n = (e_o, A^n e_o) \quad , \quad n=0,1,\ldots$$

and an associated Hamburger moment problem: one is required to find a non-decreasing function $\sigma(\lambda)$ such that

$$C_n = \int_{\infty}^{\infty} \lambda^n \, d\sigma(\lambda) \ . \tag{5}$$

If A is e.s.a. then it has a unique self-adjoint extension $\bar{A}=A^{++}$ which can be represented as

$$\bar{A} = \int_{\infty}^{\infty} \lambda \, dE(\lambda)$$

where $E(\lambda)$ is an increasing family of orthogonal projection operators. One has

$$\bar{A}^n = \int_{-\infty}^{\infty} \lambda^n \, dE(\lambda)$$

so that a solution to the moment problem is given by

$$\sigma(\lambda) = (e_o, E(\lambda)e_o) \ . \tag{6}$$

Also the solution is unique and the moment problem is said to be determinate.

On the other hand if A is not e.s.a. then it has infinitely many self-adjoint extensions and there are correspondingly infinitely many solutions to the moment problem which is then said to be indeterminate. The situation is summarised in the following theorem which is proved in Akhiezer (1965).

Theorem 1

The moment problem associated with the J-matrix A is determinate if and only if A is e.s.a.. A necessary and sufficient condition for this to occur is

$$\sum_{n=0}^{\infty} |Q_n(Z)|^2 = \infty \quad \text{for} \quad \text{Im}\, Z \neq 0 .$$

The determinateness criteria of Theorem 1 is not very useful in practice. The Carleman condition which is expressed in terms of the growth properties of the sequence

$$C_n = (e_0, A^n e_0)$$

or the off diagonal elements $b_n$ is more easily verified. For a proof of the following see Akhiezer (1965) and Masson and McClary (1972).

Theorem 2

The J-matrix A is e.s.a. if one of the following holds:

(i) $\sum_{n=1}^{\infty} \|A^n e_0\|^{-1/n} = \infty$

(ii) $\sum_{n=0}^{\infty} b_n^{-1} = \infty$

(iii) A is semibounded and $\sum_{n=1}^{\infty} \|A^n e_0\|^{-1/2n} = \infty$

(iv) A is semibounded and $\sum_{n=0}^{\infty} b_n^{-1/2} = \infty$ .

Note that A is semibounded if it is either bounded below $((f,Af) \geq \text{const.}(f,f),\ f \in \mathcal{D}(A))$ or bounded above $((f,Af) \leq \text{const.}(f,f),\ f \in \mathcal{D}(A))$.

In the case when the J-matrix is e.s.a. one can obtain useful information from the truncated matrices $A_N = P_N A P_N$ where $P_N$ projects onto the subspace spanned by the basis vectors $e_n$, $n=0,1,\ldots,N$. The effect of $P_N$ is, of course, to put $a_n=b_n=0$ for $n>N$. A direct computation shows that

$$(e_o,(I - ZA_N)^{-1} e_o) = [N/N + 1](Z) , \tag{7}$$

the $[N/N + 1]$ Padé approximant to the series

$$\sum_{n=o}^{\infty} C_n Z^n , \qquad C_n = (e_o, A^n e_o) .$$

The approximants $(I - ZA_N)^{-1}$ have convergence properties which are analogous to the convergence properties of Padé approximants (Masson, 1970).

<u>Theorem 3</u>

Let the J-matrix A be e.s.a.. Then for Im Z $\neq$ 0, $\text{s.lim.}_{N\to\infty}(I - ZA_N)^{-1} = (I - Z\bar{A})^{-1}$. In particular, $\lim_{N\to\infty}[N/N + 1](Z) = (e_o,(I - Z\bar{A})^{-1} e_o)$ . The convergence is uniform with respect to Z in a compact region which is a positive distance from the real axis.

<u>Theorem 4</u>

Let $E_N(\lambda)$ be the spectral family associated with $A_N$. Let A be e.s.a.. Then $E_N(\lambda) \underset{s}{\to} E(\lambda)$ for $\lambda$ not equal to an eigenvalue of $\bar{A}$. In particular $\sigma_N(\lambda)$, the spectral function associated with the $[N/N + 1]$ Padé approximant, converges to $\sigma(\lambda) = (e_o, E(\lambda) e_o)$ (the solution to the moment problem) for $\lambda$ not equal to an eigenvalue of $\bar{A}$.

<u>Theorem 5</u>

Let A be e.s.a.. If $f(\lambda)$ is a bounded continuous function of $\lambda \in \mathbb{R}$, then $f(A_N) \underset{s}{\rightarrow} f(\bar{A})$.

Note that the operator convergence in Theorems 3, 4 and 5 is strong convergence (convergence in norm when acting on a vector) while the Padé convergence refers to weak convergence (convergence of scalars).

## 3. Jacobi Matrices for a General Symmetric Operator

We consider the case now of a general symmetric operator A. One can generate Jacobi matrices by using what is called the method of moments (Vorobyev, 1965). Let $f \in C^\infty(A)$ where $C^\infty(A) = \bigcap_{n=1}^{\infty} \mathcal{D}(A^n)$, the set of vectors which are in the domain of $A^n$ for all n. On the space $L_f(A)$ = linear span $\{f, Af, \ldots\}$ and its completion $H_f(A) = \bar{L}_f(A) \subseteq H$ one has A represented by a Jacobi matrix provided that dim $L_f = \infty$. This becomes evident if one orthogonalises the sequence of vectors $f, Af, \ldots$ by means of the Gram-Schmidt procedure to obtain $f_o = f, f_1 = (A-a_o)f_o$ and

$$f_{n+1} = (A-a_n)f_n - b_{n-1}^2 f_{n-1}, \quad n=1,2,\ldots \quad . \tag{8}$$

This yields an orthonormal basis $e_n = f_n/\|f_n\|$ for $H_f$ with

$$\begin{aligned} a_n &= (e_n, Ae_n) \\ b_n &= (e_n, Ae_{n+1}) = \|f_{n+1}\| / \|f_n\| . \end{aligned} \tag{9}$$

With respect to this basis one has $H_f$ isomorphic to $\ell^2$ and A represented by a Jacobi matrix with Eqs. (3) and (4) a consequence of Eqs. (8) and (9).

Thus one may use the results of Section 2 to obtain information about A on the subspace $H_f(A)$. Also the piecing together of information on various subspaces $H_f(A)$ will provide one with global information about A. For this purpose it is convenient to define some special classes of $C^\infty$ vectors which we denote by $U, A, Q$ and $S$.

(i) Vectors of uniqueness:

$U = \{f |$ A is represented by an e.s.a. J-matrix in $H_f(A)\}$

(ii) Analytic vectors:

$$A = \{f | \sum_{n=0}^{\infty} \frac{\|A^n f\|}{n!} s^n < \infty, \ s > 0\}$$

(iii) Quasianalytic vectors:

$$Q = \{f | \sum_{n=1}^{\infty} \|A^n f\|^{-1/n} = \infty\}$$

(iv) Stieltjes vectors:

$$S = \{f | \sum_{n=1}^{\infty} \|A^n f\|^{-1/2n} = \infty\} .$$

It follows immediately that $A \subseteq Q \subseteq S$. Also from the Carleman condition of Theorem 2 one has $Q \subseteq U$ and $S \subseteq U$ if A is semibounded. The following theorem concerning a global property of A follows easily from the definition of $U$ and the above inclusions (Masson and McClary, 1972).

Theorem 6

Let A be a symmetric operator in a Hilbert space. Then A is e.s.a. if one of the following holds.

(i) A has a dense set of vectors of uniqueness.
(ii) A has a dense set of analytic vectors.
(iii) A has a dense set of quasianalytic vectors.
(iv) A is semibounded and has a dense set of Stieltjes vectors.

Condition (ii) was proved by Nelson (1959) using Stone's theorem on the generators of one parameter unitary groups. Conditions (i), (iii) and (iv) were first proved by Nussbaum (1965) and (1969) and alternate proofs are contained in Masson and

McClary (1972), Simon (1971) and Chernoff (1972). Quasi-analytic vectors have also been shown by Hasegawa (1971) to be important for more general types of operators such as quasi-accretive operators for which $\mathrm{Re}(f,Af) \geq \mathrm{const.}(f,f)$.

## 4. Applications

I will briefly describe three types of applications of these ideas to problems in quantum theory.

(1) Self-adjointness of Hamiltonians:

One can use the results on Jacobi matrices to prove the self-adjointness of certain Hamiltonians of quantum theory (Masson and McClary, 1971 and 1972, de Brucq and Tirapegui, 1970 and Tirapegui, 1971). The Hamiltonians that have been considered are all related to the anharmonic oscillator Hamiltonian

$$H = \frac{1}{2}\left(-\frac{d^2}{dx^2} + x^2\right) + \lambda x^{2m},\ \lambda > 0$$

which acts in the space $L^2(\mathbb{R}^1)$. This has the form of a Jacobi matrix with respect to the basis $\{\phi_n\}$ where the $\phi_n$ are eigenvectors of the "free" Hamiltonian

$$H_o = \frac{1}{2}\left(-\frac{d^2}{dx^2} + x^2\right) .$$

One uses the property

$$x\phi_n = \sqrt{\frac{n+1}{2}}\,\phi_{n+1} + \sqrt{\frac{n}{2}}\,\phi_{n-1}$$

to obtain a bound on the off diagonal elements $((\phi_n,H\phi_{n+1}) < \mathrm{const.}\ n^m)$. Theorem 2 then yields e.s.a. for $m\leq 2$. This method has been successfully applied to prove e.s.a. of the local Hamiltonian for the $\lambda(\phi^{2m})_2$, $m=1,2$ quantum field theory model in two space-time dimensions. It is not known how to treat the case $m>2$ by means of Jacobi matrices.

(2) Eigenvalue calculations (McClary, 1972 and Brändas and Micha, 1972):

The eigenvalue problem $H\psi_\lambda = E_\lambda \psi_\lambda$ for the self-adjoint operator $H = H_o + \lambda V$ with $H_o > 0$ may be cast into the following form

$$\psi_\lambda = \psi_o + \lambda H_o^{-\frac{1}{2}} (1 + A(E_\lambda))^{-1} f \qquad (10)$$

where $E_\lambda$ satisfies the Brillouin-Wigner implicit formula

$$E_\lambda = E_o + \lambda(\psi_o, V\psi_o) - \lambda^2(f, (1+A(E_\lambda))^{-1} f) \qquad (11)$$

with $A(E) = H_o^{-\frac{1}{2}} P(\lambda V-E)P\, H_o^{-\frac{1}{2}}$ and $f = H_o^{-\frac{1}{2}} PV\psi_o$ Here $H_o\psi_o = E_o\psi_o$ and P projects onto the space orthogonal to $\psi_o$. Eqs. (10) and (11) may be solved by the use of Theorem 3 which gives one an approximation to the resolvent $(1+A(E))^{-1}$. This procedure has been justified for the vacuum eigenvalue problem for the $\lambda(\phi^{2m})_2$ quantum field theory models with m=2 or 3. It is not known how to treat the case m>3 although an alternative Borel summation method has been shown to work for the anharmonic oscillator for any m.

(3) Time independent scattering theory (Baker, 1965 and Alabiso et al., 1972):

The resolvent approximation of Theorem 3 may also be used to solve Lippmann-Schwinger type equations of scattering theory.

$$Tf = \lambda Vf + \lambda VG_o Tf$$

where V is the "potential", $G_o$ the "free" Greens function and T is the T or K matrix.

(i) Case V > 0. One has the formal solution

$$Tf = \lambda \sqrt{V} (1 - \lambda A)^{-1} \sqrt{V} f$$

where $A = \sqrt{V} G_o \sqrt{V}$ is symmetric if $G_o$ is symmetric.

(ii) Case $G_o > 0$. One has the formal solution

$$Tf = \lambda Vf + \lambda^2 V\sqrt{G_o}\,(1 - \lambda A)^{-1}\,\sqrt{G_o} V\,f$$

where $A = \sqrt{G_o}\,V\sqrt{G_o}$ is symmetric.

(iii) In the case where neither V nor $G_o$ is positive one may still consider the formal solutions above provided that $\sqrt{G_o}$ or $\sqrt{V}$ exist. However the operator A is then not symmetric. The approximation $A_N = P_N A P_N$ with $P_N$ projecting onto a space generated by N powers of A is still justified if A is a compact operator (Vorobyev, 1965). The connection with the Padé approximant is however lost. If, on the other hand, one uses a projection $P_N$(non-orthogonal) associated with a biorthogonal basis generated by N powers of A and $A^+$ one recovers the Padé approximant. The convergence properties of these approximants are however not known.

## References

Akhiezer N. I., (1965). "The Classical Moment Problem", Oliver and Boyd, Edinburgh and London.

Alabiso C., Butera C. and Prosperi G. M., (1972). Nucl. Phys. B42, 493.

Baker G. A., Jr., (1965). In "Advances in Theor. Physics", (K. A. Brueckner, ed.), Vol.1, pp.1-58, Academic Press, N.Y.

Brändas E. J. and Micha D. A., (1972). J. Math. Phys., 13, 155.

Chernoff P. R., (1972). "Some Remarks on Quasi-Analytic Vectors", preprint, Univ. of California, Berkeley.

de Brucq D. and Tirapegui E., (1970). Nuovo Cimento 67A, 225.

Hasegawa N., (1971). Proc. Am. Math. Soc. 29, 81.

Masson D., (1970). In "The Padé Approximant in Theoretical Physics", (G. A. Baker and J. L. Gammel, eds.), Academic Press, N.Y.

Masson D. and McClary W. K., (1971). Comm. Math. Phys. 21, 71.

Masson D. and McClary W. K., (1972). J. Func. Anal. 10, 19.

McClary W. K., (1972). Comm. Math. Phys. 24, 171 and "The Classical Moment Problem and Quantum Field Theories in Two Space-Time Dimensions", thesis, Univ. of Toronto.

Nelson E., (1959). Ann. of Math. 70, 572.

Nussbaum A. E., (1965). Arkiv för Matematik 6, 179.

Nussbaum A. E., (1969). Studia Math. 33, 305.

Simon B., (1971). Indiana Univ. Math. J. 20, 1145.

Stone M. H., (1932). "Linear Transformations in Hilbert Space", American Math. Society, Providence, R.I.

Tirapegui E., (1971). Nuovo Cimento 3A, 561.

# ALTERNATIVE APPROACH TO PADE APPROXIMANTS

R. C. Johnson

(Mathematics Department, Durham University, England)

## 1. Introduction

This paper outlines an alternative approach to Padé Approximants, where the main ideas are developed by essentially geometrical constructions. The object is to stimulate intuition by giving a simple and concrete picture to hold in mind, and thus to provide new insights into the mechanisms of the Padé process.

There are two main sections. The first sets up three related geometrical pictures and points out some of their main features. The second deals further with a couple of suggestions immediately raised by the geometrical approach. These concern firstly a view of the Padé process as the finding of a solution to a fixed-point problem (especially relevant to physical applications), and secondly and more briefly the possibility of avoiding in part the notorious pole problem.

This discussion concentrates on the central ideas and avoids detailed technicalities.

## 2. Geometrical Pictures

The object of attention is a sequence *S* of formal approximations $S_o, S_1, \ldots, S_r, \ldots$ to some interesting quantity. The immediate example is the succession of partial sums of a power series expansion, but equally we may consider for example numerical quadrature estimates, iterative approximations to a root of an equation - indeed whatever is produced by the particular application at hand. The adjective 'formal' refers to the possibility that $S$ is not converging in the usual sense. Whatever the sequence, the aim is to estimate either its limit, or if it diverges, some useful generalisation thereof. The Padé method appears as the natural technique, as follows.

Suppose we take three terms of $S$ (say $S_r$, $S_{r+1}$, $S_{r+2}$). Then a simple way to exploit the limit information they contain is Aitken's method (e.g. Noble, 1964). That is, we form differences, $\Delta S_{r+1} = S_{r+1} - S_r$, $\Delta S_{r+2} = S_{r+2} - S_{r+1}$, and then through the points $(S_r, \Delta S_{r+1})$, $(S_{r+1}, \Delta S_{r+2})$ on a Cartesian $(S, \Delta S)$ plot draw a straight line to extrapolate to $\Delta S=0$. Fig. 1 illustrates a simple case.

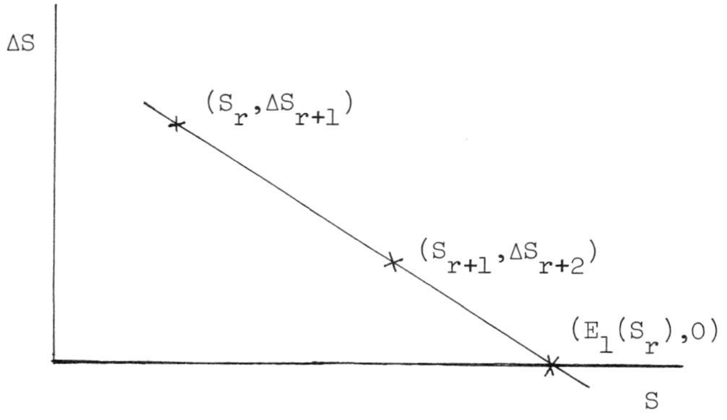


Fig. 1

The result is the point $(E_1(S_r), 0)$, where what we call $E_1(S_r)$ is given by the familiar $\delta^2$-formula:

$$E_1(S_r) = \frac{S_{r+1}^2 - S_r S_{r+2}}{2S_{r+1} - S_r - S_{r+2}} . \qquad (1)$$

Now if in fact we are dealing with partial sums $S_r$ of a power series f

$$S_r = \sum_{k=0}^{r} a_k x^k , \qquad (2)$$

$$f = a_0 + a_1 x + a_2 x^2 + \ldots , \qquad (3)$$

then we immediately discover that after cancellations the rational function $E_1(S_r)$ is exactly its $[1, 1+r]$ Padé approximant.

There are two points to note from this example:

(i) If the differences are increasing with r, (*S* tending to diverge) the extrapolation is still well-defined - in this case it gives an estimate of the apparent limit of the completely reversed sequence

$$\bar{S} = \ldots S_{r+1}, S_r, S_{r-1}, \ldots , S_1, S_0$$

based on the three terms $S_{r+1}$, $S_r$, $S_{r-1}$. This is obvious from the symmetry of eq. (1). We refer to the apparent limit of $\bar{S}$ as the "antilimit" of *S*.

(ii) A zero of the Padé denominator occurs when $\Delta S_{r+1} = \Delta S_{r+2}$, when the extrapolating line in Fig. 1 is parallel to the S-axis, and then $E_1(S_r)$ is undefined.

To use more than three terms of *S* it is possible to make nonlinear extrapolations through more points on Fig. 1, but at best this is clumsy. We choose the more elegant method of linear extrapolation of differences to zero, but in higher dimensions.

Specifically, what we call $E_n(S_r)$ is obtained by constructing from the 2n+1 terms $S_r$, $S_{r+1}, \ldots, S_{r+2n}$ the n+1 Euclidean (n+1)-vectors

$$\underline{u}^{(k)} = \begin{pmatrix} S_{r+k} \\ \Delta S_{r+k+1} \\ \Delta S_{r+k+2} \\ \cdot \\ \cdot \\ \cdot \\ \Delta S_{r+k+n} \end{pmatrix} , \quad k=0,1,\ldots,n, \qquad (4)$$

when the point

$$\underline{p} = \begin{pmatrix} E_n(S_r) \\ 0 \\ 0 \\ \cdot \\ \cdot \\ \cdot \\ 0 \end{pmatrix} \qquad (5)$$

is where their common n-dimensional hyperplane $P_n^{(U)}$ intersects the S-axis.

The natural expectation is that the quantities $E_2(S_r)$, $E_3(S_r)$,... etc. are increasingly more efficient limit estimates, since they exploit more of $S$ and involve extrapolating more and more differences to zero simultaneously.

An explicit formula for $E_n(S_r)$, and thus its connection with Padé approximants, is easily found.

If $\underline{U}$ is the normal to the extrapolating hyperplane, we have

$$\underline{U}^+\underline{u}^{(o)} = \underline{U}^+\underline{u}^{(1)} = \ldots = \underline{U}^+\underline{u}^{(n)} = \underline{U}^+\underline{p} , \qquad (6)$$

and defining

$$z_i = -U^*_{i+1}/U^*_1 , \qquad i=1,2,\ldots,n ,$$

$$z_{n+1} = E_n(S_r) , \qquad (7)$$

we re-arrange (6) to obtain the linear equations

$$\underline{Az} = \underline{C} . \quad (8)$$

Here $\underline{A}$ is (n+1)-square:

$$\underline{A} = \begin{pmatrix} \Delta S_{r+1} & \Delta S_{r+2} & \cdots & \Delta S_{r+n} & 1 \\ \Delta S_{r+2} & \Delta S_{r+3} & \cdots & \Delta S_{r+n+1} & 1 \\ \vdots & & & \vdots & \vdots \\ \Delta S_{r+n+1} & \cdots & & \Delta S_{r+2n} & 1 \end{pmatrix} , \quad (9)$$

and $\underline{C}$ is an (n+1)-vector:

$$\underline{C} = \begin{pmatrix} S_r \\ S_{r+1} \\ S_{r+2} \\ \vdots \\ S_{r+n} \end{pmatrix} \quad (10)$$

Cramer's rule then gives

$$E_n(S_r) = \det \underline{A}^c / \det \underline{A} , \quad (11)$$

where $\underline{A}^c$ has as its last column the vector $\underline{C}$, but otherwise is identical to $\underline{A}$. (For example, for n=1, $E_1(S_r)$ is a ratio of 2×2 determinants, expanding to give eq. (1)).

From eq. (11) we immediately recognise $E_n(S_r)$ as being in fact what is known as the ($n^{th}$-order) epsilon transform of the sequence $S_r, S_{r+1}, \ldots, S_{r+2n}$, usually written as $e_n(S_{r+n})$, (Shanks, 1955).

And therefore if we are dealing with successive partial sums of a power series, the rational function $E_n(S_r)$ is (after cancellation) precisely its $[n,n+r]$ Padé approximant.

This allows us to write the ultra-compact formula

$$[n,n+r] = (\underline{A}^{-1}\underline{C})_{n+1} , \tag{12}$$

where for r<0 we need to define $S_r=0$. (This is 'ultra-compact' because it is neater than Baker's (1971) generalisation of "Nuttall's compact formula"). Note that increasing n corresponds to following a paradiagonal of the Padé Table, (i.e. a line parallel to the principal diagonal).

So in a simple and illustrative way the epsilon transform appears as a natural higher-order generalisation of Aitken's acceleration method, and Padé approximants arise from its application to a sequence of partial sums.

Further insight is obtained by introducing two more, similar, geometrical constructions.

The first of these we call the V-picture (to distinguish it from the preceding U-picture) where the vectors

$$\underline{v}^{(k)} = \begin{pmatrix} S_{r+k} \\ S_{r+k+1} \\ S_{r+k+2} \\ \cdot \\ \cdot \\ \cdot \\ S_{r+k+n} \end{pmatrix} , \quad k=0,1,\ldots,n \tag{13}$$

have a common hyperplane $P_n^{(V)}$ (normal $\underline{V}$) passing through

$$\underline{q} = \begin{pmatrix} 1 \\ 1 \\ 1 \\ \cdot \\ \cdot \\ \cdot \\ 1 \end{pmatrix} \times E_n(S_r) . \tag{14}$$

This follows directly by a 'rotation' $\underline{R}$ of the U-picture

$$\underline{v}^{(k)} = \underline{Ru}^{(k)} , \qquad \underline{q} = \underline{Rp} ,$$

where $R_{ij}=1$ for $j \leq i$, and $R_{ij}=0$ for $j>i$. From eq. (6) the normals transform as $\underline{V}=(\underline{R}^{-1})^T\underline{U}$.

Fig. 2 illustrates the V-picture for n=1 in a simple case.

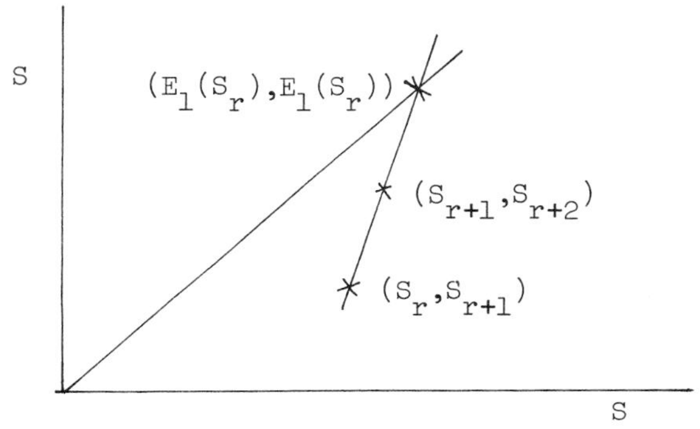


Fig. 2

Evidently $E_1(S_r)$ estimates the limit or antilimit as an eventual point of equality between successive terms, forwards or backwards, in the sequence.

This feature generalises to all n, and is the reason for looking at the V-picture.

For by inspection $E_n(S_r)$ is invariant under the complete reversal

$$(S_r, S_{r+1},\ldots,S_{r+2n})$$

$$\rightarrow (S_{r+2n}, S_{r+2n-1},\ldots,S_{r+1}, S_r) .$$

(This is true of course in the U-picture (etc.), but is less obvious.)

It may be helpful at this point to see in this context how Padé deals with a rank-m geometric sequence

$$S_r = a + \sum_{i=1}^{m} b_i \rho_i^r \quad , \qquad (\rho_i \neq 1, \forall\ i) \ ,$$

where it is well known that $E_n(S_r) = a$ for $n+r \geq m$.

This result is understood simply by observing that in such a sequence any m+1 successive terms are linearly related, thus defining precisely the V-picture hyperplane for $n+r \geq m$. In fact we find immediately that for $k=0,1,\ldots,n$, $n \geq m$, $r \geq 0$

$$S_{r+m+k} - \left(\sum_{i=1}^{m} \rho_i\right) S_{r+m+k-1} + \left(\sum_{i>j} \rho_i \rho_j\right) S_{r+m+k-2} - \cdots$$

$$\cdots + (-1)^m \left(\prod_{i=1}^{m} \rho_i\right) S_{r+k} = a\left(\prod_{i=1}^{m} (1-\rho_i)\right) \ , \quad (15)$$

where the components of the normal $\underline{V}$ with normalisation $\Sigma V_i^* = \pi(1-\rho_i)$ appear as the coefficients on the left-hand side.

Dividing out by $\Sigma V_i^*$, we see at once the invariance under the operation of sequence-reversal if in the correspondingly re-ordered components of $\underline{V}$ we change $\rho_i \to 1/\rho_i$. That is, if convergence and divergence are interchanged.

This illustrates the quite general result that as far as Padé is concerned, limit and antilimit of $S$ are of precisely equal significance. Both can be regarded in some way as a point of equality between successive terms in a somehow interpolated and continued sequence. The next section takes this further.

For a third geometrical construction, we form higher differences $\Delta^2 S_i = \Delta S_i - \Delta S_{i-1}$ etc., and then the vectors

$$\underline{w}^{(k)} = \begin{pmatrix} S_{r+k} \\ \Delta S_{r+k+1} \\ \Delta^2 S_{r+k+2} \\ \cdot \\ \cdot \\ \cdot \\ \Delta^n S_{r+k+n} \end{pmatrix}, \quad k=0,1,\ldots,n , \qquad (16)$$

which define a hyperplane $P_n^{(W)}$ (normal $\underline{W}$) also passing through the U-picture point $\underline{p}$ (eq. (5)) on the S-axis.

Again, simply 'rotate'

$$\underline{w}^{(k)} = \underline{Rv}^{(k)} , \quad \underline{p} = \underline{Rq} , \quad \underline{W} = (\underline{R}^{-1})^T \underline{V} , \qquad (17)$$

where now $R_{ij}=0$ for $j>i$, $R_{ij}=1$ for $i=j$, and for $j<i$ we have

$$R_{ij} = \frac{(-1)^{i+j}(i-1)!}{(i-j-2)!(j+1)!} \qquad (18)$$

The point of this W-picture is that the Padé process is shown as one of simultaneous extrapolations to zero of all differences up to the $n^{th}$. This reinforces the U-picture expectation that as n increases $E_n(S_r)$ is a progressively better (anti)limit estimate.

To summarise so far: each picture has its useful aspect. The natural approach is through U or W pictures which give the standard formulae easily and emphasise why $E_n(S_r)$ can be expected to be more efficient the larger is n. In conventional Padé terms they lead us to expect that $[N,M]$ estimates the sum or 'antisum' more closely as $N,M\to\infty$ with N-M fixed. The V-picture on the other hand shows naturally the unity of limit and antilimit - an aspect deserving deeper mathematical study.

Note however the common feature of each picture - the occurence of poles (zeros of the Padé denominator, det $\underline{A}$, eq. (11)) when the extrapolating hyperplane lies parallel to the vector it is supposed to intersect. In this situation the hyperplane itself generally remains well defined, but there is the separate possibility of its disappearance, as discussed below.

We consider in turn two interesting suggestions of the geometrical approach; first a view of an (anti)limit of $S$ as a solution to a fixed point problem, and second a hint about pole avoidance. Neither of these aspects is yet fully developed - they both seem to offer promising lines of research.

## 3. Fixed Points

The V-picture introduces the Padé limit and antilimit of a sequence on manifestly equal footing, suggesting their unification somehow as "a point of equality of successive terms". This implies setting up a unique smooth interpolation and extrapolation of the sequence.

A suitable interpolation can be devised in a natural way if successive terms of $S$ are related by a specific recurrence rule, for then we may introduce what we call the generator, g, of $S$.

The generator is defined as the function(al) that builds $S$ iteratively from the first term $S_o$ according to

$$S_{j+1} = g(S_j) . \qquad (19)$$

As an example of a generator of a sequence of partial sums of a power series, consider the Gauss hypergeometric function $F(\alpha,\beta;\gamma;x)$, with expansion

$$1 + \frac{\alpha\beta}{\gamma}\, x + \frac{\alpha(\alpha+1)\ \beta(\beta+1)}{\gamma(\gamma+1)}\, \frac{x^2}{2!} + \dots . \qquad (20)$$

Many common functions are special cases.

The generator here is

$$g(S) = 1 + xS(x) + (\alpha+\beta-\gamma-1) \int_0^x S(x)\, dx$$

$$+ (\alpha-\gamma)(\beta-\gamma) \int_0^x x^{-\gamma} \int_0^x x^{\gamma-1} S(x) dx\, dx, \quad (21)$$

with $S_0=1$.

This is one example of a large class of series where successive expansion coefficients $a_m$ are related by

$$\frac{a_{m+1}}{a_m} = \text{(rational function of m)} ,$$

and where therefore g can be written explicitly as a combination of integral and differential operators. Further examples are easy to construct. In all cases, when g is applied to a polynomial of degree r (not just a partial sum) it produces one of degree r+1. For purely numerical sequences, g is simply a function.

This hypergeometric example, and others below, illustrate how g characterises $S$, containing any "smoothness" information relevant for attempted extrapolations beyond a few given terms. We note that in many applications of Padé approximants the generator g is in fact the object primarily available - for example we may try to sum with Padé approximants a quantum mechanical Neumann-type perturbation expansion, where the generator is just the integral equation of scattering that gives partial sums by the Born iteration.

With the generator idea established, the next step is the immediate identification of the (anti)limit of $S$ as a solution $\sigma$ to

$$S = g(S) . \quad (22)$$

This is suggested as a more precise statement of the hitherto rather vague limit and antilimit ideas. For if $S$ is converging, g contracts, and iteration gives $\sigma$ as the usual limit. Otherwise g still exists, and (22) may still have a solution appearing as "the point from which $S$ diverges", or as the limit of the reversed sequence $\bar{S}$ - i.e. the antilimit of $S$. Both limit and antilimit (if either (or both?) exist) are where successive terms of $S$, extrapolated through g, become equal.

As a simple example, for the geometric series $1+x+x^2+\ldots$ we have $g(S)=1+xS$ so that for all $x\neq 1$ the generalised sum in this sense is $\sigma=(1-x)^{-1}$, (which of course in this case Padé gives exactly). Likewise, for the pathologically divergent series $1-1!x+2!x^2-3!x^3\ldots$, we have $g(S)=1-x\frac{d}{dx}(xS(x))$, and the appropriate solution $S=\sigma$ to the differential equation $S=1-x(xS)'$ is

$$\sigma(x) = \int_0^\infty \frac{e^{-t}dt}{1+tx} \quad .$$

Again, this is the well known Padé sum of the series.

In any case, arguing from the V-picture geometrical construction and supported by such examples, it appears that $E_n(S_r)$ extrapolates towards $\sigma$, and so we are led to the following conjecture: that the Padé process systematically estimates fixed points of the generator.

As further support for this idea we note first that indeed, as required, every fixed point of g is generally (i.e. subject to det $\underline{A}\neq 0$) a fixed point of $E_n$. For, if $S_r$ approaches $\sigma$ then via g we have $S_{r+1}\to S_{r+2}\to\ldots\to S_{r+2n}\to\sigma$, and the linear equations $\underline{Az}=\underline{C}$ of (8) become n+1 statements that

$$E_n(\sigma) = \sigma \quad . \tag{23}$$

The details of the approach to $\sigma$ depend on the nature of g, and determine whether $E_n$ contracts and whether $E_n(S_r)$ converges to $\sigma$ as n increases. The W-picture, of simultaneous extrapolations to zero of all differences from first to $n^{th}$, suggests that this is so if g is sufficiently well behaved.

As a model, consider a numerical sequence $S$ where the terms are produced by a generating function that is expandable about the point $S=\sigma$:

$$g(\sigma+\epsilon) = \sigma + c_0\epsilon + \ldots , \quad c_0 \neq 1 \quad . \qquad (24)$$

Then if $S_r=\sigma+\epsilon$, succeeding terms $S_{r+1}$, $S_{r+2}$ ... are geometrically related to $O(\epsilon^2)$, and so we find easily that

$$E_n(\sigma+\epsilon) = \sigma + c_n\epsilon^{n+1} + \ldots \quad . \qquad (25)$$

Thus proceeding in this way for this smooth generator and for $|\epsilon| = |S_r-\sigma|$ small enough, we discover that $E_n$ contracts and $E_n(S_r)\to\sigma$ as $n\to\infty$.

The rate of convergence in the model depends on the size of $|c_0|$, for the coefficients $c_n$ etc. in (25) involve the factor $c_0^n/(c_0^n-1)$. (The case $c_0=1$ has to be treated separately.) Numerical experiments invariably show very rapid convergence and for example confirm the prediction that if g has two or more fixed points convergence is preferentially towards that where $|c_0|$ is smallest.

We conclude therefore that there is good encouragement for more serious investigation of these ideas. The current task is to examine individual generators, or classes of generators, to try to see if convergence proofs are more easily or more generally formulated this way than by more conventional methods.

## 4. Catastrophes

The immediately preceding discussion ignores possible poles of $E_n(S_r)$ - zeros of det $\underline{A}$. These are usually what severely limit convergence proofs. However, the present geometrical approach offers a way to sidestep at least one class of such catastrophes.

The point here is that we see the mechanism of the singularities explicitly, and recognise that they are of two types.

The first class, which can be circumvented easily, is when the extrapolating hyperplane slips sideways to lie parallel to the vector it is supposed to intersect. To avoid these class one catastrophes, we simply shift attention from the approximant $E_n(S_r)$ to the hyperplane itself as n increases.

Class two catastrophes may still occur, however, when there is a loss of the hyperplane - it loses one or more dimensions if the differences between the defining vectors happen to be linearly dependent. Then $\det \underline{A} = \det \underline{A}^c = 0$.

To be more specific, we know that $P_n$ is defined by its normal and one intersecting vector. In say the V-picture (normal $\underline{V}$) we may choose the vector $\underline{v}^{(o)} \equiv \underline{C}$ (eq. (10)). Then with the normalisation

$$\sum_{i=1}^{n+1} V_i = 1$$

we have from (6) the (super-ultra-compact) formula (c.f. eq. (15))

$$E_n(S_r) = \underline{V}^+\underline{C} . \qquad (26)$$

A class one catastrophe is signalled by loss of normalisation, $\Sigma V_i = 0$, but both $\underline{V}$ and $\underline{C}$ generally remain finite.

A recurrence scheme for $\underline{V}$ is not hard to devise, giving essentially the over-square matrix taking the n-vector normal associated with $E_{n-1}(S_r)$ into the (n+1)-vector for $E_n(S_r)$. A class two catastrophe appears here as the inability to construct the recurrence matrix.

We arrive here at a point of current development, which is a good place to stop the discussion.

## 5. Conclusions

It is important to re-emphasise the object of this approach to Padé approximants - that is, to give at least some definite geometrical pictures of how they work in order to aid the intuition, and more generally to stimulate new ideas by offering some different perspectives.

This presentation has omitted any reference to the points usually emphasised, such as projective invariance properties, unitarity, etc., of Padé approximants, and not mentioned the special status of series of Stieltjes, nor for example the links with variational methods. Also, non-paradiagonal sequences have been neglected. It is rewarding to look at these and other such standard topics from the present viewpoint.

Indeed, even though the speculations and suggestions made in Section 3 and 4 may ultimately turn out to be completely misguided, their discussion and investigation in every possible way is certainly profitable.

## References

Baker G. A., (1971). In "The Padé Approximant in Theoretical Physics" (G. A. Baker and J. L. Gammel, eds.), pp. 1-39, Academic Press, N.Y.

Shanks D., (1955). J. Math. and Phys. (MIT), 34, pp. 1-42.

Noble B., (1964). "Numerical Methods", pp. 33-36, Oliver and Boyd.

# NONLINEAR PADÉ APPROXIMANTS FOR LEGENDRE SERIES

J. Fleischer

(Los Alamos Scientific Laboratory, University of California

Los Alamos, New Mexico 87544, U. S. A.)

## 1. Introduction

Padé approximants (PA's) for Legendre series have been introduced recently by Holdeman, J. T., Jr. (1969) and Fleischer, J. (1972). These were obtained by "cross-multiplication" which yields a linear system of equations for their computation--as is the case for PA's for Taylor series. However, "linear" Legendre PA's do not have the property that their first expansion coefficients agree with the first coefficients of the original series. If one wants to define PA's for Legendre series which have this property, one has to solve transcendental equations and we call these PA's "nonlinear" Padé approximants.

## 2. The nonlinear approximants

The definition of Padé approximants for Taylor series can be made as

$$f(z)Q_M(z) - P_N(z) = A\, z^{M+N+1} + \ldots \tag{1}$$

or equivalently

$$f(z) - \frac{P_N(z)}{Q_M(z)} = A'\, z^{M+N+1} + \ldots \quad , \tag{2}$$

where $P_N$ and $Q_M$ are polynomials of degree $N$ and $M$, respectively. Introducing PA's for Legendre series analogous definitions lead to different approximations with completely different properties. The reason is that in the expansion of a product of Legendre polynomials more than one term contributes:

$$P_i(z)P_k(z) = \sum_{\ell=|i-k|}^{i+k} \alpha_\ell^{(i,k)} P_\ell(z) \quad . \tag{3}$$

The definition analogous to equation (1) leads to the "linear"approximants, while the definition analogous to equation (2), with which we are concerned in this paper, reads:

$$f(z) - \frac{R_N(z)}{S_M(z)} \equiv f(z) - \frac{\sum_{i=1}^{N} n_i P_i}{\sum_{k=1}^{M} d_k P_k} = \bar{a}_{M+N+1} P_{M+N+1} + \dots \quad . \tag{4}$$

Putting $d_0 = 1$, we expand $1/S_M$ in a Legendre series:

$$\frac{1}{S_M} = \frac{1}{d_0P_0 + d_1P_1 + \dots + d_MP_M} \equiv \frac{s_0}{(z-s_1)(z-s_2)\dots(z-s_M)}$$

$$= D_0P_0 + D_1P_1 + \dots + D_{M-1}P_{M-1} + \dots \quad .$$

Performing the projection into partial waves, we have:

$$D_\ell = -(2\ell+1) \sum_{\nu=1}^{M} \frac{P_\ell(s_\nu)}{S_M'(s_\nu)} Q_0(s_\nu) \quad , \quad \ell = 0, \dots, M-1 \quad ,$$

where we assumed that the roots $s_\nu$ of the polynomial $S_M$ are not degenerate, $Q_0$ being the Legendre function of the second kind and index 0. We obtain the higher coefficients recursively:

$$D_M = \frac{2M+1}{d_M} \left(1 - \sum_{\lambda=0}^{M-1} \frac{d_\lambda}{2\lambda+1} D_\lambda\right)$$

and

$$D_{M+\nu} = - \frac{1}{d_M \alpha_\nu^{(M,M+\nu)}} \sum_{\lambda=\lambda_0}^{M+\nu-1} \left( \sum_{\mu=0}^{M} d_\mu \alpha_\nu^{(\mu,\lambda)} \right) D_\lambda \quad , (\nu = 1,2,3, \ldots)$$

with

$$\lambda_0 = \mathrm{Max}(0, \nu - M) \quad ,$$

$\alpha_\nu^{(\mu,\lambda)}$ being the coefficients in equation (3). Equation (4) now reads:

$$(n_0 P_0 + n_1 P_1 + \ldots n_N P_N)(D_0 P_0 + D_1 P_1 + \ldots + D_{M-1} P_{M-1} + \ldots)$$

$$= a_0 P_0 + a_1 P_1 + \ldots a_{M+N} P_{M+N} + \ldots \quad .$$

Using (3) and equating the coefficients of Legendre polynomials with equal index on both sides of this equation, we finally obtain the following system of equations for the coefficients $n_i (i = 0, \ldots, N)$ and $d_i (i = 1, \ldots, M)$:

$$\sum_{\nu=0}^{N} n_\nu \sum_{\mu=|\lambda-\nu|}^{\lambda+\nu} D_\mu \alpha_\lambda^{(\mu,\nu)} = a_\lambda \quad , \lambda = 0, \ldots, N+M \qquad (5)$$

This is a linear system of equations for the $n_\nu (\lambda = 0,1, \ldots, N)$ and a transcendental system for the $d_\mu (\lambda = N+1, \ldots, N+M)$. Solutions have been found by means of the fitting routine $FIT_4$, developed by W. Anderson and T. Doyle. In table 1 we compare various approximations for the generating function

$$f(z) = \frac{1}{\sqrt{1 - 2az + a^2}} = \sum_{\ell=0}^{\infty} a^\ell P_\ell(z) \quad ,$$

where $a = 0.3$ has been chosen.

| z | PS | LP | NP | exact |
|---|---|---|---|---|
| - 3.5 | - | .561 | .5600 | .5599 |
| - 1.75 | .77 | .6838 | .683591 | .683586 |
| - 1.0 | .76940 | .769236 | .76923081 | .76923077 |
| - .5 | .84824 | .8481884 | .84818895 | .84818893 |
| - .0 | .95781 | .957829 | .95782626 | .95782629 |
| + .5 | 1.12505 | 1.125080 | 1.12508792 | 1.12508790 |
| + 1.0 | 1.4283 | 1.42849 | 1.4285709 | 1.4285714 |
| + 1.5 | 2.13 | 2.26 | 2.290 | 2.294 |
| + 1.75 | 2.82 | 3.19 | 4.52 | 5.00 |

Table 1. PS is the partial sum ($l_{max} = 6$), LP the linear [2/2] - PA, NP the nonlinear [3/3] - PA--both involving the same number of coefficients--and the exact function.

Quite generally we observe that the nonlinear PA's are a better approximation than the linear ones and that their improvement of convergence is quite remarkable. Even when the Legendre series diverges ($z = -3.5$) they still give an extremely good result.

We remark that the system (5) has more than one solution in general, but only one solution with all the poles outside the region $-1 \leqslant z \leqslant +1$ has been found in the cases under consideration. Furthermore, in the above case the poles of the solution were situated on the cut, separated by zeroes--as is expected for PA's in general. We finally remark that the following theorem can be proved (Fleischer, J. (1972)):

Let $Q_k(z)$ be any infinite sequence of [N/M] PA's to a formal Legendre series where M + N tends to infinity with k. If the absolute value of the $Q_k$ is uniformly bounded in the ellipse (its boundary included) with foci at +1 and - 1 and semi-major axis A, then the $Q_k$ converge uniformly in the ellipse with semi-major axis a, a < A to an analytic function f(z), the Legendre series of which has a semi-major axis of at least A.

## 3. Concluding remarks

We have shown that it is possible to find solutions of the system of transcendental equations, which determines the coefficients of the nonlinear PA's. Though the search for the solution is a fairly involved procedure, particularly for higher order approximants, the method may become a useful tool for the summation of Legendre series in many cases of physical interest.

## References

Work performed under the auspices of the United States Atomic Energy Commission and the North Atlantic Treaty Organization.

Fleischer, J. (1972). Nucl. Phys. B 37, 59.

Fleischer, J. (1972). Los Alamos preprint (LA-DC-72-347).

Holdeman, J. T., Jr. (1969). Mathematics of Computation 23, 275.

# ORTHOGONAL POLYNOMIALS AND LACUNARY APPROXIMANTS

F. V. Atkinson
(Mathematics Department, University of Toronto, Canada)

## 1. Introduction

Let $s(x)$, $a \leq x \leq b$, be real, non-decreasing, with an infinity of points of increase, and define

$$F(z) = \int_a^b ds(x)/(z - x). \tag{1}$$

It is then appropriate to seek Padé approximants in terms of descending powers of $z$, where $z$ is large. We ask for an n-th degree polynomial $p_n(z) = z^n + \ldots$ such that $p_n(z)F(z)$ has a Laurent expansion in which some given sequence of n coefficients must all vanish. In the main case we ask that, for large $z$,

$$p_n(z)F(z) = q_n(z) + O(z^{-n-1}), \tag{2}$$

where $q_n(z)$ is a polynomial of degree n-1. Then (see e.g. Szegö (1948), Wall (1948), Wynn (1960) we have :

(i) $p_n(z)$ is the n-th degree orthogonal polynomial determined by $p_n(z) = z^n + \ldots$ together with

$$\int_a^b p_n(x)x^m \, ds(x) = 0 \, , \, m = 0, \ldots, n-1, \tag{3}$$

(ii) $q_n(z)/p_n(z)$ has n simple poles in $(a,b)$, with positive residues there which are the "Cotes-Christoffel numbers" associated with the "mechanical quadrature",

(iii) for large $z$, $q_n(z)/p_n(z) \to F(z)$ as $n \to \infty$.

In what follows I outline an extension to polynomials subjected to side-conditions of certain types; one such side-condition might require the derivatives of the polynomials to vanish on a prescribed point-set.

## 2. Approximation and orthogonality

One may view (2) not only as an approximation, but also as a form of orthogonality; we are asking that $p_n(z)F(z)$ be orthogonal to $z^m$, $m = 0, \ldots, n-1$ in the sense that the integral of the product around a large circle should vanish. More generally, let $f_m(z)$, $m = 0, 1, \ldots,$ be entire functions, and suppose that

$$p_n(z) = f_n(z) + a_{n,n-1}f_{n-1}(z) + \ldots + a_{n,o}f_o(z) \tag{4}$$

satisfies the orthogonality requirements on $(a,b)$ that

$$\int_a^b p_n(x)f_m(x)ds(x) = 0 \;, \; m = 0, \ldots, n-1; \tag{5}$$

we have then the second set of orthogonality conditions

$$\int p_n(z)F(z)f_m(z)\, dz = 0, \; m = 0, \ldots, n-1 \;. \tag{6}$$

the contour being any sufficiently large circle.

## 3. The modified polynomials

We apply the above to the system of polynomials

$$f_o(x) = 1, \; f_m(x) = \int_o^x t^{m-1}\theta(t)dt, \quad m \geqslant 1, \tag{7}$$

where $\theta(x)$ is some fixed polynomial, with real coefficients, of degree k say, which is positive in $[a,b]$; the standard case is given by $k = 0$, and so we suppose $k \geqslant 1$. Since (7) forms a Tschebyshev system in $[a,b]$, we have that $p_n(x)$ as determined by (4), (5) will have just n simple zeros in $[a,b]$, indeed in $(a,b)$. In view of its degree it will have, if $n \geqslant 1$, a further k "exterior" zeros (Atkinson 1954, 1955), which tend to those of $\theta(x)$ as $n \to \infty$.

## 4. The modified approximation

We introduce the rational functions

$$g_m(z) = (z^{m+1}/\theta(z))' \tag{8}$$

and now ask, in place of (2), that

$$p_n(z)F(z) = \Sigma_{-\infty}^{n+2k-1} b_m g_m(z) \tag{9}$$

for large z , where

$$b_m = 0 \ , \ m = -2, \ldots, -n \ ; \tag{10}$$

here we pass over the cases $n = 0, 1$ . It is easily checked that (4), (9), (10) ensure (6) and (5).

Equivalently, one may ask that

$$F(z) = Q_n(z)/P_n(z) + O(z^{-2n-1-2k}) \tag{11}$$

where

$$P_n(z) = \int^z \theta(w)U_n(w)dw \ , \quad Q_n(z) = (V_n(z)/\theta(z))', \tag{12}$$

and $U_n$ , $V_n$ are polynomials of degrees $n-1$, $n+2k-1$ .

We now sketch the argument leading to properties similar to (ii), (iii) of §1. We must show that the "interior" zeros of $P_n(z)$ (or $p_n(z)$), those in (a,b), are separated by zeros of $Q_n(z)$; more precisely, one may show that, at least for large n, the residues of $Q_n(z)/P_n(z)$ at its poles in (a,b) are positive. As in the standard case, this may be discussed by applying the "mechanical quadrature" to a suitable polynomial; here it must have degree less than 2n+2k, be positive or zero in (a,b), vanishing at all but one of the interior zeros of $P_n$ and at all of its exterior zeros, and have derivative containing $\theta(z)$ as a factor. Some special cases were discussed in Atkinson (1954).

One may now show that $Q_n(z)/P_n(z) \to F(z)$ as $n \to \infty$, uniformly outside a circle, centre 0 , which contains in its interior $[a,b]$ and all zeros of $\theta(z)$ . One ingredient in the proof is the fact that, by (11), each coefficient in the Laurent expansion of $Q_n(z)/P_n(z)$ is fixed for all large n. We note further that

$$Q_n(z)/P_n(z) = (s(b) - s(a))z^{-1} R_n(z)S_n(z),$$

where $R_n(z)$ , $S_n(z)$ are rational functions which tend to 1 as $z \to \infty$, $R_n(z)$ incorporates all linear factors arising from the interior zeros of $P_n$ and $Q_n$ , while $S_n(z)$ comprises linear factors associated with all other zeros or poles of $Q_n/P_n$ , which are bounded in number as $n \to \infty$ . We can use also the fact that the exterior zeros of $P_n(z)$ this is less obvious, and is a by-product of the proof.

## 5. Specializations and extensions.

The term "lacunary" is prompted by the case (Atkinson, 1954), $\theta(x) = x^k$ , $0 < a < b$. Here the polynomials (7) are marked by the absence of the powers $x, \ldots, x^k$ . The same powers are absent in the denominators of the associated approximants (11).

The requirement that $\theta(x)$ be positive in $[a,b]$ can be relaxed to that of positivity in $(a,b)$. For a discussion of the behavious of the exterior zeros in a case of this nature I refer to Atkinson (1955).

It seems that similar developments may be possible for polynomials generated by sets of the form

$$1, \int_0^x \theta_1(x_1)dx_1, \int_0^x \theta_1(x_1)dx_1 \int_0^{x_1} \theta_2(x_2)dx_2, \ldots$$

where $\theta_m(x)$ , $m = 1, 2, \ldots$, satisfy the above restrictions on $\theta(x)$. We get the standard case on taking all the $\theta_m(x)$ to be 1, the above case on taking all but the first to be 1. A special case would be given by the polynomials generated by an arbitrary subset of the powers of x.

## REFERENCES

Atkinson F.V., (1954), Revista (Tucumán), Ser. A. Mat. y Fis.Teor. 10, 95

Atkinson F.V., (1955), Monatshefte f. Math. 59, 323

Szegö G., (1948), "Orthogonal Polynomials" (New York)

Wall H.S., (1948), "Analytic Theory of Continued Fractions" (New York)

Wynn P., (1960), Math. Comp. 14, 147

NUMERICAL ANALYSIS AND NUMERICAL METHODS

# RECURSIVE CALCULATION OF PADE APPROXIMANTS*

George A. Baker, Jr.

(Applied Mathematics Department, Brookhaven National Laboratory, Upton, N.Y. 11973, U.S.A.)

There are two main ways in which Padé approximants have been calculated. The first is the direct solution of the Padé equations for the $[L/M]$ Padé approximant defined as

$$[L/M] = P_L(x)/Q_M(x) \tag{1}$$

where $P_L(x)$ is a polynomial of degree L and $Q_M(x)$, a polynomial of degree M. When we approximate a formal power series

$$f(x) = \sum_{j=0}^{\infty} f_j \, x^j \tag{2}$$

the explicit equations are

$$Q_M(x)\, f(x) - P_L(x) = O(x^{L+M+1}) \tag{3}$$

$$Q_M(0) = 1.0 \; . \tag{4}$$

As long as the determinant

$$C(L/M) = \det \begin{vmatrix} f_{L-M+1} & \cdots & f_L \\ \vdots & \ddots & \vdots \\ f_L & \cdots & f_{L+M-1} \end{vmatrix} \tag{5}$$

does not vanish, there is a unique solution to Eq. (3) and (4). If none of the C(L/M) vanish, then the Padé table is said to be <u>normal</u>. In this case, much more efficient methods of obtaining results can be given based on the structure of the Padé equations.

Most of the fundamental identities required are due to Frobenius (1881), with one further important one added by Wynn (1966). We will not derive them here, but instead indicate how this derivation can be accomplished and what sort of identities can be expected.

First we have the two-term identities. Since by Eq. (3),

$$f(x) - P_L^{(J)}/Q_M^{(J)} = O(x^{L+M+1})$$

$$f(x) - P_{L+1}^{(J)}/Q_{M+1}^{(J)} = O(x^{L+M+3}) , \qquad (6)$$

where we define J=L-M for the [L/M] Padé approximant. If we subtract these equations and multiply the result by $Q_M^{(J)}\ Q_{M+1}^{(J)}$ we obtain

$$P_{L+1}^{(J)}\ Q_M^{(J)} - P_L^{(J)}\ Q_{M+1}^{(J)} = O(x^{L+M+1}) . \qquad (7)$$

However, as the left hand side of (7) is a polynomial of degree at most (L+M+1), we see that (7) yields just the single term $x^{L+M+1}$. One can deduce by Sylvester's determinant identity that

$$\frac{P_{L+1}^{(J)}(x)}{Q_{M+1}^{(J)}(x)} - \frac{P_L^{(J)}(x)}{Q_M^{(J)}(x)} = \frac{[C(L+1/M+1)]^2\ x^{L+M+1}}{Q_M^{(J)}(x)\ Q_{M+1}^{(J)}(x)} \qquad (8)$$

where we use the normalisation $Q_M^{(J)}(0)=C(M+J/M)$. Similarly, one can deduce other identities for other adjacent entries in the Padé table, provided we are not troubled by vanishing C(L/M)'s.

Next (Baker, 1970) we can apply the two term identities to obtain the cross-ratio identities. A cross-ratio is a ratio of the form

$$\frac{(Z_1-Z_2)(Z_3-Z_4)}{(Z_1-Z_3)(Z_2-Z_4)} = R . \qquad (9)$$

Each of the $Z_i$ appears in the numerator and the denominator. If we let the $Z_i$ represent four adjacent entries in the Padé table, then by identities of the form (8) R becomes a monomial in x over the product of the four Q's divided by a denominator of the same structure. As the four Q's cancel, we find R to be just a constant times a simple power (0 or 1 usually) of x. We can use these cross ratio results to derive directly the three term Padé identities. First solve (9) for $Z_1$ as

$$Z_1 = \frac{Z_2(Z_3-Z_4) - Z_3\ R(Z_2-Z_4)}{(Z_3-Z_4) - R(Z_2-Z_4)} \quad . \tag{10}$$

If we now substitute Padé approximants for the $Z_i$, Eq. (10) becomes

$$\frac{P_1}{Q_1} = \frac{P_2(P_3Q_4-Q_3P_4) - P_3\ R(P_2Q_4-P_4Q_2)}{Q_2(P_3Q_4-Q_3P_4) - Q_3\ R(P_2Q_4-P_4Q_2)} \quad . \tag{11}$$

Now by the two term identities, the terms in parentheses are reduced to simple form. When we adjust (11) for normalisation by multiplying numerator and denominator by a common factor, we deduce that the $P_i$ and $Q_i$ both satisfy a three term recursion relation with simple coefficients. If $S_i$=a $P_i$+b $Q_i$, then

$$S_1 = \alpha(x)\ S_2 + \beta(x)\ S_3 \tag{12}$$

where, for most adjacent triples, the $\alpha$ and $\beta$ are L and M dependent constants or linear polynomials.

By explicit evaluation of the coefficients in the three term identities (12) and by using a combination of them we can deduce Frobenius's star or quadratic identity

$$S(L+1/M)\ S(L-1/M) - S(L/M+1)\ S(L/M-1) = S(L/M)^2 \tag{13}$$

which reduces to a special case of Sylvester's determinant identity for $S_i=Q_i$ and x=0. We can also deduce in the same way Wynn's identity

$$\frac{1}{[L+1/M] - [L/M]} + \frac{1}{[L-1/M] - [L/M]}$$

$$= \frac{1}{[L/M+1] - [L/M]} + \frac{1}{[L/M-1] - [L/M]} \qquad (14)$$

which, like the Frobenius one, involves five terms arranged in a star shaped pattern in the Padé table. This last one has the advantage that it relates the values of the Padé approximants directly without reference to the series coefficients.

With these identities in mind, we are now in a position to discuss the recursive calculation of Padé approximants. There are really three distinct problems. First, to compute the value of a Padé approximant at a specific point. Second, to compute the Padé approximant itself as a function of x. Third, to compute the limiting locations of the poles and zeros of the Padé approximant.

The best solution to the first or value problem seems to us to be the improved version of the ε-algorithm based on Wynn's identity (14). The pattern of the work is

$$\begin{array}{cccccccc}
 & 0 & & 0 & & 0 & & \\
\infty & [0/0] & & [0/1] & & [0/2] & & [0/3] \\
 & \downarrow & \nearrow & & \nearrow & & \nearrow & \\
\infty & [1/0] & & [1/1] & & [1/2] & & \\
 & \downarrow & \nearrow & & \nearrow & & & \\
\infty & [2/0] & & [2/1] & & & & \\
 & \downarrow & \nearrow & & & & & \\
 & [3/0] & & & & & &
\end{array} \qquad (15)$$

where the $[L/0]$ are just the partial sums of the Taylor series, and using the boundary conditions shown in (15), we can by following the arrows solve (14) successively and easily for any other Padé approximants in the upper left triangular part of the table.

The best solution to the second or approximant problem seems to us Baker's algorithm (1970).** We introduce the sequence

$$\frac{\eta_{2j}(x)}{\theta_{2j}(x)} = [N-j/j] , \quad \frac{\eta_{2j+1}(x)}{\theta_{2j+1}(x)} = [N-j-1/j] . \quad (16)$$

If we evaluate the coefficients of the three term identities (12) for successive triples in the sequence, and express them in terms of the coefficients of the Padé approximants themselves, we get the recursion relations

$$\frac{\eta_{2j}(x)}{\theta_{2j}(x)} = \frac{(\bar{\eta}_{2j-1}\,\eta_{2j-2}(x) - x\,\bar{\eta}_{2j-2}\,\eta_{2j-1}(x))/\bar{\eta}_{2j-1}}{(\bar{\eta}_{2j-1}\,\theta_{2j-2}(x) - x\,\bar{\eta}_{2j-2}\,\theta_{2j-1}(x))/\bar{\eta}_{2j-1}}$$

$$\frac{\eta_{2j+1}(x)}{\theta_{2j+1}(x)} = \frac{(\bar{\eta}_{2j}\,\eta_{2j-1}(x) - \bar{\eta}_{2j-1}\,\eta_{2j}(x))/(\bar{\eta}_{2j}-\bar{\eta}_{2j-1})}{(\bar{\eta}_{2j}\,\theta_{2j-1}(x) - \bar{\eta}_{2j-1}\,\theta_{2j}(x))/(\bar{\eta}_{2j}-\bar{\eta}_{2j-1})} \quad (17)$$

where $\bar{\eta}_j$ is the coefficient of the highest power of x in $\eta_j(x)$ (i.e., $x^{n - [\frac{1}{2}(j+1)]}$). The starting values for these recursion relations are given by

$$\eta_o(x) = \sum_{k=o}^{n} f_k x^k , \quad \theta_o(x) = 1.0 ,$$

$$\eta_1(x) = \sum_{k=o}^{n-1} f_k x^k , \quad \theta_1(x) = 1.0 . \quad (18)$$

The division factors in (17) of $(\bar{\eta}_{2j-1})$ and $(\bar{\eta}_{2j} - \bar{\eta}_{2j-1})$ are simply to maintain the normalisation condition $\theta_j(0) = 1.0$. The procedure of Longman (1971) is similar, but uses the second recursion formula (17) to generate the whole upper left triangular portion of the table.

The third or root problem is solved by Rutishauser's (1954) Q.D. algorithm. We will give Gragg's variant (1971). These procedures, suffice it to say, are to be employed when the poles or zeros of vertical or horizontal sequences in the Padé tables tend to a limit. This situation would clearly be true for meromorphic functions in which no two poles lie at the same distance from the origin for vertical sequences, or zeros for horizontal ones. More elaborate discussions can be given when these stipulations do not apply.

For the case when they do apply, Hadamard (1892) proved that

$$\lim_{L\to\infty} \frac{C(L+1/M-1)\ C(L-1/M)}{C(L/M-1)\ C(L/M)} = u_M \tag{19}$$

where $u_M$ is the location of the $M^{th}$ pole of f(x) in order of distance from the origin. The quantities which appear in (19) are just the ratio of first and last coefficients in the denominators of the [L/M-1] and [L-1/M] Padé approximants and so (19) gives the location of the extra pole in the [L/M] sequence over those in the [L/M-1] sequence. By Hadamard's theorem for determinants of reciprocal series in terms of determinants for the direct series we can use (19) to give results for the zeros of the function as well. They are

$$\lim_{M\to\infty} \frac{C(L-1/M+1)\ C(L/M-1)}{C(L-1/M)\ C(L/M)} = v_L\ . \tag{20}$$

The Q.D. algorithm can be set up in terms of the u-v table and the rhombus rules. If we define

$$u(L/M) = \frac{C(L+1/M-1)\ C(L-1/M)}{C(L/M-1)\ C(L/M)}$$

$$v(L/M) = \frac{C(L-1/M+1)\ C(L/M-1)}{C(L-1/M)\ C(L/M)} \tag{21}$$

then it can be shown that

$$u(L/M)\ v(L/M) = u(L/M+1)\ v(L+1/M) \tag{22}$$

$$u(L/M+1) + v(L+1/M) = u(L+1/M+1) + v(L+1/M+1) \tag{23}$$

which are called the rhombus rules. Equation (22) can be verified directly from the definitions and Eq. (23) by Sylvester's determinant identity.

We lay out the work in a u-v table as

$$\begin{array}{cccccccccccc}
 & 0 & & & & 0 & & & & 0 & & \\
0 & & & v(1/1) & & & & v(1/2) & & & & v(1/3) \\
 & & \nearrow & & & & \nearrow & & & & \nearrow & \\
 & u(1/1) & & & & u(1/2) & & & \nearrow & u(1/3) & & \\
 & & & & \nearrow & & \searrow & & & & & \\
0 & \downarrow & & v(2/1) & & (23) & & v(2/2) & & & & \\
 & & \nearrow & & \searrow & & \nearrow & & & & & \\
 & u(2/1) & & (22) & & u(2/2) & & & & & & \\
 & & \searrow & & \nearrow & & & & & & & \\
0 & \downarrow & & v(3/1) & & & & & & & & \\
 & & \nearrow & & & & & & & & & \\
 & u(3/1) & & & & & & & & & &
\end{array} \tag{24}$$

where the border values are as shown, and we may initialise with

$$u(L/1) = f_{L-1}/f_L \tag{25}$$

and then use, as appropriate, either Eq. (22) or (23) to generate successive vertical rows of the u-v table. Plainly the solution alternately of (22) and (23) is either a quotient, or a difference, which is where the name comes from. It is to be noted that each calculation involves the entries at the corners of a rhombus.

One can equally well start with the v's as initial values, instead of the u's. The initial values of the v's are

$$v(1/M) = d_{M-1}/d_M \tag{26}$$

where

$$d(x) = 1/f(x) = \sum_{m=o}^{\infty} d_j\ x^j \ . \tag{27}$$

As an illustration we have worked out a few rows of the u-v table for

$$f(x) = (1-3x + 2x^2)/(2-3x)$$

$$d(x) = 2 + 3x + 5x^2 + 9x^3 + 17x^4 + 33x^5 + 65x^6 + \dots \quad (28)$$

The table begins

$$\begin{array}{ccccccccccccc}
 & 0 & & 0 & & 0 & & 0 & & 0 & & & \\
0 & & \frac{2}{3} & & \frac{3}{5} & & \frac{5}{9} & & \frac{9}{17} & & \frac{17}{33} & \dots & \to \frac{1}{2} \\
 & -\frac{2}{3} & & \frac{1}{15} & & \frac{2}{45} & & \frac{4}{153} & & \frac{8}{561} & & \dots & \\
0 & & -\frac{20}{3} & & \frac{9}{10} & & \frac{17}{18} & & \frac{33}{34} & & & \dots & \to 1
\end{array} \quad (29)$$

Convergence of the v's to the roots of the numerator of (28) is illustrated.

*Work supported in part by the U.S.A.E.C. and in part by A.R.P.A. through the Materials Science Centre of Cornell University.

**See also the article by P. J. S. Watson in these proceedings for an equally good, new algorithm.

## References

Baker G. A., Jr., (1970). In "The Padé Approximant in Theoretical Physics" (G. A. Baker and J. L. Gammel, eds.), pp.1-39, Academic Press, N.Y.

Frobenius G., (1881). Jour. für Math. (Crelle) 90, 1.

Gragg W. B., (1972). SIAM Review 14, 1.

Hadamard J., (1892). Jour. de Math. (4) 8, 101.

Longman I. M., (1971). Inter. J. Computer Math. (B) 3, 53.

Rutishauser H., (1954). Z. Angew. Math. Phys. 5, 233.

Wynn P., (1966). Numer. Math. 8, 264.

# ALGORITHMS FOR DIFFERENTIATION AND INTEGRATION

P.J.S. Watson

(Physics Department, Carleton University, Ottawa, Canada)

Abstract

We discuss various algorithms which enable Padé approximants and continued fractions to be used for numerical integration and differentiation of functions. The emphasis is on numerical experimentation rather than formal results.

The uses of Padé approximants (P.A.) and continued fractions (C.F.) in various special problems in mathematical physics is well known. It is of interest to ask whether it is practical to use them as a standard technique for solving numerical problems which arise in practice, and we have carried out some numerical experiments to see whether it can be used as a practical tool for the integration of arbitrary functions.

The integration of a Taylor series is clearly trivial. It is probably true that the large majority functions arising in physical problems consist of combinations of functions whose series expansions are well known, so we may assume that most integrals in practice have an expansion of the form

$$f(x) = x^{\nu} \sum_{n=o}^{\infty} a_n x^n \quad , \ \nu \ \text{real} \tag{1}$$

Hence the problem really reduces to manipulating series, which can be done fairly efficiently by computer: the routines we have written handle any combination of binomials, trigonometric, exponential and Bessel functions, and it is possible to consider functions of functions (e.g. $J_{1/3}(\sin(x))$ ) and inverse functions (e.g. arctan (x) ). The resulting series must then be converted into a C.F. or P.A. The former may be accomplished by the following algorithm : write

$$\frac{g_0 x^{\nu}}{1+} \quad \frac{g_1 x}{1+} \quad \frac{g_2 x}{1+} \cdots = \sum_{k=1}^{\infty} b_k \, x^{k+\nu} \qquad (2)$$

and define

$$c_k^{(o)} = b_k \; ; \quad d_o^{(o)} = 1 \text{ when } k = 0$$
$$d_k^{(o)} = 0, \text{when } k \neq 0 \qquad (3)$$

We may then calculate the $g_i$ sequentially by

$$g_i = \frac{c_o^{(i)}}{d_o^{(i)}} \quad , \quad d_k^{(i+1)} = c_k^{(i)}$$
$$c_k^{(i+i)} = g_i d_k^{(i)} - c_{k+1}^{(i)} \qquad (4)$$

As is well known, it is possible to derive the P.A. from the C.F. via the recurrence relation

$$Q_{n+1}(x) = Q_n(x) + g_n \, Q_{n-1}(x) \qquad (5)$$

and this gives rise to a direct algorithm for calculating the P.A. Write

$$Q_n(x) = \sum_{m=o}^{\infty} q_m^{(n)} \; x^m \qquad (6)$$

and it is possible to derive

$$g_m = - \frac{\Sigma \, q_k^{(m)} \; b_{m+1-k}}{\Sigma \, q_k^{(m-1)} b_{m-k}} \qquad (7)$$

where the summation is over the non-zero values of $q_k^{(m)}$. This appears to be a new algorithm for the evaluation of the P.A. coefficients. Numerically this method is a little worse than Baker's algorithm (Baker (1971)) but requires less storage space; it would be convenient if rational arithmetic were available.

We compare some typical results with 8 and 24 point Gaussian integration.

| INTEGRAL | ABSOLUTE ERRORS | | |
|---|---|---|---|
| | P.A. | 8 point | 24 point |
| $\int_{-0.9}^{0} \frac{dx}{1+x}$ | $4.10^{-8}$ | $1.10^{-4}$ | $1.10^{-11}$ |
| $\int_{0}^{10} \frac{\sin x}{x}\, dx$ | $< 10^{-12}$ | $5.10^{-10}$ | $< 10^{-12}$ |
| $\int_{0}^{1} \frac{dx}{\sqrt{-\log x}}$ | $2.10^{-5}$ | $1.10^{-1}$ | $4.10^{-2}$ |
| $\int_{0}^{1} \operatorname{arc\,sin} x\, dx$ | $1.10^{-6}$ | $5.10^{-7}$ | $< 10^{-10}$ |

More extreme examples are given by $\int_0^x \sec^2 y\, dy$, where it is hard to imagine any conventional method of integration working for $x > {}^{\pi}/2$: the P.A. has an error of less than $10^{-6}$ at $x = 4.5$.

The P.A. is catastrophically wrong for $\int_0^1 (2 + \sin 10\, x)^{-1} dx$, due presumably to an infinite set of branch cuts lying near the real axis, while Gaussian quadrature gives an error of about $10^{-6}$.

The method can be straightforwardly extended to differential equations and even integral transforms. For Laplace transforms we consider

$$f(z) = s^{\nu} \sum_{n=0}^{\infty} a_n\, s^n$$

Then

$$F(t) = \int_0^{\infty} f(s)\, e^{-st}\, dt = \sum_{n=0}^{\infty} a_n \frac{\Gamma(n+\nu+1)}{t^{n+\nu+1}}$$

and the series in $^1/_t$ can be converted into a P.A. again. The results are about as satisfactory as those for the simple integrals

This method relies on analytic knowledge of the function. From numerical knowledge of the function f(x), we can construct a C.F. of the second kind, y(x) such that $y(x_i)=f(x_i)$ $(i=o\ldots N)$

$$y(x) = c_o + \frac{x-x_o}{c_1+} \quad \frac{x-x_1}{c_2+} \quad \frac{x-x_2}{c_3+} \quad \ldots\ldots \tag{9}$$

which may be converted into a P.A. by the usual recurrence relation. It is then possible to integrate the P.A. itself; however the process appears to be rather unstable. It is interesting to note that there is an algorithm for the <u>differentiation</u> of y(x) which offers some advantages over conventional methods for numerical differentiation.

We can evaluate the C.F. at any point x by the recurrence relation

$$y_k = \frac{x-x_{k-1}}{c_k+y_{k+1}} = \frac{u}{v} \tag{10}$$

where $y_N=0$ and the interpolated value is given by $y_o$. This relation can be differentiated to yield

$$y_k' = \frac{v-uy'_{k+1}}{v^2} \tag{11}$$

where $y_N' = 0$ and the interpolated value for the derivative is simply $y_o'$ : note that the recursion relation depends on $y_{k+1}$ as well as $y'_{k+1}$. Successive differentiation of (11) may be performed at the cost of steadily increasing complexity : we have considered derivatives of up to $4^{th}$ order. Below are shown the region for which the relative error < $10^{-6}$ for two functions; in each case we have used as input 20 points lying between 0 and 4.

Limits of error < $10^{-6}$

| f(x) | = | sin x | | log x | |
|---|---|---|---|---|---|
| g(x) | | -1.7 | 6.04 | .24 | > 9.5 |
| $g^{1}(x)$ | | -1.22 | 6.63 | .23 | 8.89 |
| $g^{11}(x)$ | | -1.04 | 5.52 | .34 | 6.96 |
| $g^{111}(x)$ | | - . 7 | 4.61 | .36 | 6.01 |
| $g^{(1V)}(x)$ | | - .21 | 4.87 | .48 | 5.46 |

## REFERENCES

G.A. Baker, 1971, Chap. 1 in "The Padé Approximant in Theoretical Physics" (G.A. Baker and J.L. Gammel, eds.) Academic Press, N.Y.

# NUMERICAL DETECTION OF THE BEST PADE APPROXIMANT AND DETERMINATION OF THE FOURIER COEFFICIENTS OF THE INSUFFICIENTLY SAMPLED FUNCTIONS

Jacek GILEWICZ

(Centre de Physique Théorique, Marseille, France)

In the numerical experiments one has a finite set of the computed Taylor coefficients $c_j$ of the power series $\sum_0^\infty c_j z^j$ which represents the function f(z). We develop the purely numerical tests for the analysis of the set $\{c_j\}$. To do this we firstly must extract from $\{c_j\}$ the set of correctly computed coefficients. Secondly, we will detect the numbers M and N of the best Padé approximant $f_{NM}$. If $M \geqslant N$ it is important to know the difference $K = M - N$ corresponding to the degree of the residual polynomial in $f_{NM}$ and indicating the diagonal in the Padé table. Next we applied this method to the calculation of the Fourier coefficients. In numerical practice the Fourier (Taylor-) coefficients are obtained by integration. The number of coefficients which can be computed is bounded by the sampling of the function f(z) (Shannon theorem for the information transfer). In physical problems one often knows certain continuity properties of f(z) but we cannot express those explicitly. Consequently we have, besides the sampling, still information left on f(z). Fortunately the first coefficients are well computed and we can construct by means of them the best Padé approximant. The numerical examples show that the extrapolation of other Fourier coefficients by means of this Padé approximant gives good results. Consequently the method described above allows, in the calculation of $c_j$ , to go beyond the bound imposed by the sampling of the function.

Numerical detection of the best Padé approximant by the analysis of the Taylor coefficients

The Padé approximant of f(z) :

$$f_{NM}(z) = P_M(z)/Q_N(z) = (p_o+\ldots+p_M z^M)/(1+\ldots+q_N z^N) \qquad (1)$$

corresponding to $\sum_0^\infty c_j z^j$ ($c_o = 0$) is defined (for N = 0) by :

$$-p_k + \sum_{j=1}^{N} c_{k-j}\, q_j = -c_k \qquad k=0,1,\ldots,M \qquad (2')$$

$$\sum_{j=1}^{N} c_{k-j}\, q_j = -c_k \qquad k = M+1,\ldots,M+N \; (c_{n<0} \equiv 0) \qquad (2'')$$

The triangular part of the Padé table defined by M + N < L will be called "$P_L$-triangle" ; the matrix of the determinants $c_{NM}$ of the system (2") is called "c-table".

Theorem : "The formal power series $\sum_o^\infty c_j z^j$ is exactly reduced to the Padé approximant $f_{NM}$ (N > 0) if there exists a recurrence relation :

$$-c_k = \sum_{j=1}^{N} q_j\, c_{k-j} \qquad (3)$$

between the coefficients $c_n$ for all k > M and $q_N \neq 0$."

The form $f_{NM}(z)$ can be decomposed as follows :

$$f_{NM}(z) = R_K(z) + \sum_{i=1}^{N} \alpha_i/(\beta_i - z) \qquad (4)$$

in which $R_K$ is a polynomial of degree $K = M - N \geqslant 0$; if $K < 0$ , $R_K \equiv 0$. Therefore the coefficients $c_j$ can be expressed :

$$c_j = r_j + \sum_{i=1}^{N} \alpha_i/\beta_i^{j+1} \qquad (5)$$

We define the punctual function ρ : $\rho(n) = c_n/c_{n+1}$ (6)

(6) gives : $\lim_{n\to\infty} \rho_1(n) = \beta_1$ (7)

so $\alpha_1$ can be determined : $$\alpha_1 = \lim_{n\to\infty} (c_n \, \beta_1^{n+1}) \qquad (8)$$

This procedure enables us to detect the fundamental mode and the corresponding geometrical series. If we subtract the geometrical series from the power series, we get a new power series, which will be subjected again to the procedures (7) and (8). The speed of the convergence of $\rho_1$ in (7) depends on the radial distance $|\beta_{i+1}| - |\beta_i|$. The coefficients $c_n$ computed by numerical integration methods are all the more erroneous as the index n is larger. We extract the finite set of the well computed coefficients from the complete computed set $\{c_j\}$ by means of the following statement :

"If a function known by its numerically computed Taylor coefficients can be represented by a Padé approximant, then :

1/ $\rho(n)$ tends to a monotonic punctual function,

2/ if this monotong, once observed, is definitely broken from L, then the coefficients with the index greater than L are badly computed."

Now our problem is to detect the best element of $P_L$-triangle.

Detection of N. We resume the operations (7) and (8) N times.

Detection of polynomial $R_K$ (z) for the case M > N (K = M - N)

In practice it is preferable to estimate K before the detection of N. We notice that if $R_K(z) = r_K z^K$ , only these two values : $\rho(K - 1)$ and $\rho(K)$ may be affected (fig. 1). The irregular behaviour of $\rho(n)$ at the beginning indicates the presence of a polynomial.

Remarks on the $\rho$-method

In the case of the presence of the polynomial $R_K$, the detection of N begins from $\rho(K + 1)$. This detection is all the more difficult as K is larger, because the information on the distant poles contained in $c_n$ decreases with increasing index n. For all the N poles to be detected, one must have $c_{NM} = 0$. Unfortunately, when K is large and/or when $(|\beta_N| - |\beta_1|)$ is great, then the matrix of the system (2") is more singular and $c_{NM} = 0$. In the c-table theory, this would give the reduction of $f_{NM}$. So, from the numerical point of view, the theorems of Gragg (1972) for the c-table would lead to wrong conclusions, especially if $M \gg N$ and/or if the difference between the absolute values of the extreme poles of the function f(z) is great. It is the main reason in favour of the $\rho$-method which permits the successive detection of the poles.

The principal roles of the described method of detection of the best Pade approximant is above all the determination of the indices M and N, but not the computation of $r_n$, $\alpha_n$ and $\beta_n$. M and N being obtained, we can compute $f_{NM}(z)$ by (2).

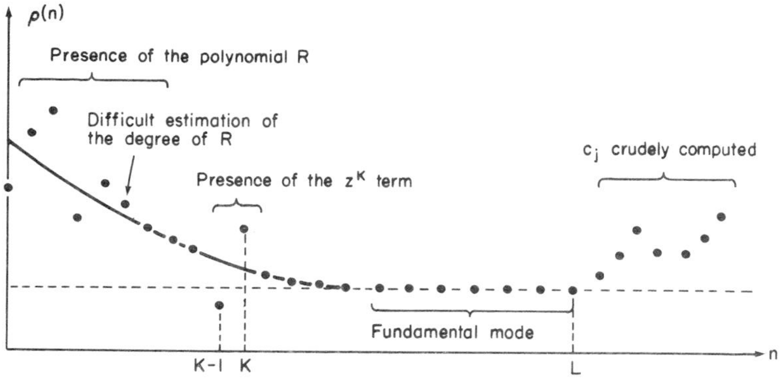


Fig. 1

Method of determination of the Fourier coefficients by the Padé approximants in the case where the function is given by the insufficient sampling for the application of the classical integral methods.

Let $g(z) = \sum_{-\infty}^{\infty} c_n z^n = \sum_{-\infty}^{-1} + \sum_{0}^{\infty} c_n z^n$ where

$$c_n = (2\pi)^{-1} \int_{-\pi}^{\pi} g(e^{ix})\, e^{-inx}\, dx.$$

On the unit circle $z = e^{ix}$ we have $f(x) = g(e^{ix}) = \sum_{-\infty}^{\infty} c_n\, e^{inx}$

We can determine for $\sum_{-\infty}^{-1}$ and $\sum_{0}^{\infty}$ of $g(z)$ the best Padé approximants by the $\rho$-method :

$$g(z) \simeq \frac{1}{z} f_{N_1M_1}\left(\frac{1}{z}\right) + f_{N_2M_2}(z) \qquad (9)$$

Other Fourier coefficients $c_j$ can be extrapolated from $f_{N_1M_1}$ and $f_{N_2M_2}$ by (3).

Particular Case. Let $g(z)$ be analytic in the unit disk and $\overline{g(z)} = g(\overline{z})$. Then $g(z)$ reduces to a sum of non-negative powers and the coefficients are real : $c_n = \pi^{-1} \int_0^{\pi} \mathrm{Re}(f(x)e^{-inx})\, dx$. We detect $f_{NM}$ from the first coefficients which have been computed correctly. Then we extrapolate $f_{NM}$ to any Fourier coefficient by (3).

Example : $g(z) = \mathrm{Log}(1.5 + z)$ and $f(x)$ sampled at 8 points on $(0,\pi)$. The trapezoidal integration gives only 14 coefficients. Our extrapolation by $f_{66}$ gives 21 ones with relative error smaller than 1%.

REFERENCE

Gragg, W.B., (1972), S.I.A.M. Review 14, 1.

APPLICATIONS OF THE ε-ALGORITHM TO QUADRATURE PROBLEMS

A. C. Genz

(Mathematical Institute, University of Kent, England)

## 1. Introduction and basic properties of the ε-algorithm

Various extrapolation, or sequence acceleration, techniques are currently used to accelerate the convergence of sequences of approximations to integrals. The now standard Romberg (1955) method, for example, and its generalisations (Joyce, 1971 reviews these) have been used effectively for large classes of integrands, but until recently the application of a powerful Padé approximant based extrapolation device, the ε-algorithm, has not been thoroughly investigated. It is the purpose of this paper to discuss the range of applicability of the ε-algorithm to quadrature problems. But before beginning the direct discussion of this application it will be useful to list some of the well known properties of the ε-algorithm and to prove an additional one.

We define the ε-algorithm (Wynn, 1956) for a given sequence of numbers $S_0, S_1, S_2, \ldots$ by the relations

$$\varepsilon_0^{(j)} = S_j \; , \quad \varepsilon_{-1}^{(j)} = 0$$

$$\varepsilon_{k+1}^{(j)} = \varepsilon_{k-1}^{(j+1)} + (\varepsilon_k^{(j+1)} - \varepsilon_k^{(j)})^{-1} \qquad (1)$$

$$j=0,1,2,\ldots \; ; \quad k=0,1,2,\ldots$$

The double array of terms generated by the ε-algorithm is usually displayed in the following form

$$\begin{array}{ccccc}
\varepsilon_{-1}^{(0)} & & & & \\
 & \varepsilon_{0}^{(0)} & & & \\
\varepsilon_{-1}^{(1)} & & \varepsilon_{1}^{(0)} & & \\
 & \varepsilon_{0}^{(1)} & & \varepsilon_{2}^{(0)} & \\
\varepsilon_{-1}^{(2)} & & \varepsilon_{1}^{(1)} & & \varepsilon_{3}^{(0)} \\
 & \varepsilon_{0}^{(2)} & & \varepsilon_{2}^{(1)} & \cdot \quad \cdot \\
\varepsilon_{-1}^{(3)} & & \varepsilon_{1}^{(2)} & \cdot & \cdot \qquad \cdot \\
 & \varepsilon_{0}^{(3)} & & \cdot & \cdot \qquad\quad \cdot \\
\varepsilon_{-1}^{(4)} & \cdot & \cdot & \cdot & \\
 & \cdot & \cdot & & \\
\cdot & \cdot & \cdot & & \\
\cdot & \cdot & & & \\
\cdot & & & &
\end{array}$$

which is called the ε-array.

If the terms $S_j$ are partial sums of a formal power series given by

$$f(x) = \sum_{i=o}^{\infty} a_i x^i, \text{ with } S_j = \sum_{i=o}^{j} a_i x_o^i \text{ for some } x_o$$

then it can be shown (Wynn, 1956 and Shanks, 1955) that

$$\varepsilon_{2k}^{(j)} = [k,k+j]_{f(x)} (x_o) \tag{2}$$

where $[k,k+j]_{f(x)} (x_o)$ is the Padé approximant to f(x) in the form

$$[k,k+j](x) = \frac{p_o+p_1x+\ldots+p_{k+j}x^{k+j}}{1+q_1x+\ldots+q_kx^k} \tag{3}$$

evaluated at $x=x_o$. So the ε-algorithm can at least be seen as an efficient computational method for computing Padé approximants for some fixed point.

When $\{S_j\}$ takes on other well-defined forms more can be said about $\varepsilon_{2k}^{(j)}$. The form which is important for quadrature applications is given by

$$S_j = S + \sum_{i=1}^{p} b_i\beta_i^j , \quad \text{with } |\beta_1|>|\beta_2|>\ldots>|\beta_p| . \tag{4}$$

If we apply the ε-algorithm to sequences whose terms are given by equation (4) then we have (Shanks, 1955)

$$\varepsilon_{2p}^{(o)} = S .$$

That is to say, given only the first 2p+1 terms of the sequence $\{S_j\}$, we can compute the limit S (or if $|\beta_\ell|>1$ for some $\ell$, then the antilimit) by the ε-algorithm. If p is large and we do not have available all of the 2p+1 terms of $\{S_j\}$ needed to compute S exactly, then we can still obtain reasonable approximations to S, assuming $|\beta_\ell|<<1$ for $\ell$ large, by the ε-algorithm because it can be shown (Wynn, 1966) that

$$\varepsilon_{2k}^{(j)} \sim S + \frac{b_{k+1}\{(\beta_{k+1}-\beta_1)\ldots(\beta_{k+1}-\beta_k)\}^2}{\{(1-\beta_1)\ldots(1-\beta_k)\}^2}\,\beta_{k+1}^j . \qquad (5)$$

A generalisation of the form given in equation (4) which is also important for the analysis of certain sequences of quadrature approximations is given by

$$S_j = S + \sum_{i=1}^{n} b_i^{(j)}\,\beta_i^j \qquad (6)$$

where $b_i^{(j)} \equiv b_i^{(j+k_0)}$, for $k_0$ some fixed integer. We can show that ε-algorithm will also extrapolate sequences in the form given by equation (6). This is done in a standard way by forming the somewhat artificial function f(λ) given by

$$f(\lambda) = S_o + (S_1-S_o)\lambda + (S_2-S_1)\lambda^2 + \ldots \qquad (7)$$

Notice that the $j^{th}$ partial sum of f(λ) for λ=1 is $S_j$, so that by applying the ε-algorithm to $\{S_j\}$ we are forming Padé approximants to f(λ) at the point λ=1. Considering f(λ) in more detail, we have

$$f(\lambda) = S + \sum_{i=1}^{p} b_i^{(o)} + \lambda \sum_{i=1}^{p} (\beta_i b_i^{(1)} - b_1^{(o)}) + \lambda^2 \sum_{i=1}^{p} \beta_i (\beta_i b_i^{(2)} - b_i^{(1)}) + \dots$$

$$= S + \sum_{i=1}^{p} b_i^{(o)} + \lambda \sum_{j=0}^{\infty} \lambda^j \sum_{i=1}^{p} \beta_i^j (\beta_i b_i^{(j+1)} - b_i^{(j)})$$

$$= S + \sum_{i=1}^{p} b_i^{(o)} + \lambda \sum_{i=1}^{p} \sum_{j=0}^{\infty} (\lambda\beta_i)^j (\beta_i b_i^{(j+1)} - b_i^{(j)})$$

$$= S + \sum_{i=1}^{p} b_i^{(o)} + \lambda \sum_{i=1}^{p} \sum_{\ell=0}^{\infty} (\lambda\beta_i)^{k_o \ell} \sum_{m=0}^{k_o-1} (\lambda\beta_i)^m (\beta_i b_i^{(k_o \ell+m+1)} - b_i^{(k_o \ell+m)})$$

$$= S + \sum_{i=1}^{p} b_i^{(o)} + \lambda \sum_{i=1}^{p} \left[ \sum_{m=0}^{k_o-1} (\lambda\beta_i)^m (\beta_i b_i^{(m+1)} - b_i^{(m)}) \right] / \left[1-(\lambda\beta_i)^{k_o}\right]$$

$$= S + \sum_{i=1}^{p} \Big[ b_i^{(o)} (1-(\lambda\beta_i)^{k_o}) + (\lambda\beta_i b_i^{(1)} - \lambda b_i^{(o)}) + \dots$$

$$+ \lambda^{k_o} \beta_i^{k_o} b_i^{(o)} - \lambda^{k_o} \beta_i^{k_o-1} b_i^{(k_o-1)} \Big] / \left[1-(\lambda\beta_i)^{k_o}\right]$$

$$= S + (1-\lambda) \sum_{i=1}^{p} \left[ \sum_{m=0}^{k_o-1} b_i^{(m)} \beta_i^m \right] / \left[1-(\lambda\beta_i)^{k_o}\right] \tag{8}$$

The right hand side of this equation can easily be written as a ratio of two polynomials in $\lambda$ with $f(\lambda=1)=S$. Because of the uniqueness of the Padé approximants, and because here the $\varepsilon$-algorithm computes Padé approximants to $f(\lambda)$ at $\lambda=1$, we expect generally from the form of equation (8) that

$$\varepsilon_{2k_o p}^{(o)} = S \ .$$

Therefore the $\varepsilon$-algorithm can be used to extrapolate sequences given by the form in equation (6). It is not difficult to show that the $\varepsilon$-algorithm also extrapolates sequences whose forms are combinations of those forms in equations (4) and (6) if enough terms of $\{S_j\}$ are available.

## 2. Application to finite interval quadrature problems

In order to apply the above information we must first obtain sequences of which we hope to accelerate the convergence. Without loss of generality we can standardise our problem by considering integrals of the form

$$If = \int_0^1 f(x)\, dx . \tag{9}$$

Given a fixed quadrature rule Qf defined by

$$Qf = \sum_{i=1}^{n} w_i\, f(x_i) \;, \quad \sum_{i=1}^{n} w_i = 1 \tag{10}$$

we will form sequences of approximations to If in a fairly standard way by successively subdividing the integration interval [0,1] into 1,2,3,... equal width subintervals and applying Qf (appropriately transformed) to each subinterval. The approximations we obtain from this process are given by

$$Q^{(m)}f = (m{\times}Q)f = \sum_{k=0}^{m-1} \sum_{j=1}^{n} \frac{w_i}{m}\, f\left(\frac{x_i+k}{m}\right) \;, \quad m=1,2,... \tag{11}$$

Under a fairly wide range of conditions on f(x), we know (see Davis and Rabinowitz, 1967) that

$$\lim_{m\to\infty} Q^{(m)}f = If .$$

What we are interested in is in what way this convergence occurs given a particular class of function f(x). If f(x) is analytic in [0,1], then a simple generalisation of the Euler-Maclaurin expansion (Lyness and Ninham, 1967) shows that $Q^{(m)}f$ is given by

$$Q^{(m)}f = If + a_1h + a_2h^2 + ... + a_ph^p + O(h^{p+1}) \tag{12}$$

with $h={}^1/m$.

This form is simplified somewhat for certain specific choices of Q, but the important point about the general form of equation (12) in terms of extrapolation techniques is that the $a_i$ are independent of m. Given a sequence $\{Q^{(m)}f\}$ one can try to eliminate the low order terms in h by combining successive terms in the sequence in various ways (Joyce, 1971) to obtain better approximations to If. In order to apply the ε-algorithm to this problem we must choose our sequence $\{S_j\}$ from $\{Q^{(m)}f\}$ in the way indicated by

$$S_j = Q^{(2^j)}f \ .$$

If this choice is made, we see that (12) can be rewritten as

$$S_j = If + a_1\beta_1^j + a_2\beta_2^j + \ldots + a_p\beta_p^j + O(\beta_{p+1}^j) \qquad (13)$$

with $\beta_i = 2^{-i}$ .

Ignoring the term $O(\beta_{p+1}^j)$, this expression is in the same form as the similar expression in equation (4), so we expect the ε-algorithm to work as an extrapolation method. The linear Romberg technique and its generalisations are more efficient for this class of problem given the usual choices of Q (Chisholm, Genz and Rowlands, 1972), however.

If f(x) has end-point singularities, the ε-algorithm can also be directly applied to a specific sequence of quadrature approximations $\{S_j\}$. Because it is difficult to describe all cases at once we will choose an important example and let f(x) be of the form

$$f(x) = x^{\alpha}(1-x)^{\gamma}\ g(x)$$

where g(x) is analytic in [0,1] and α and γ are not integers. Now $Q^{(m)}f$ has the form given by (Lyness and Ninham, 1967)

$$\begin{aligned} Q^{(m)}f = If &+ c_1h^{1+\alpha} + c_2h^{2+\alpha} + \ldots + c_ph^{p+\alpha} + O(h^{p+\alpha+1}) \\ &+ d_1h^{1+\gamma} + d_2h^{2+\gamma} + \ldots + d_ph^{p+\gamma} + O(h^{p+\gamma+1}). \end{aligned} \qquad (14)$$

Although the $c_i$ and $d_i$ are still independent of h it will be more difficult to develop general linear extrapolation methods for equation (14) because each different combination of $\alpha$ and $\gamma$ will have to be separately treated (Fox, 1967). But by choosing the m sequence to be 1,2,4,... and ignoring the order terms we have for $S_j$

$$S_j \equiv Q^{(2^j)}f = If + \sum_{i=1}^{p} c_i\beta_i^j + \sum_{i=1}^{p} d_i\delta_i^j \qquad (15)$$

with $\beta_i = 2^{-(\alpha+1)}$ , $\delta_i = 2^{-(\gamma+i)}$ .

This form is clearly amenable to ε-algorithm extrapolation without taking any special account of the fact that $\alpha$ and $\gamma$ are not integers. Many other types of endpoint singularity, including the important logarithmic case, will produce approximation sequences in a form similar to that in equation (14), so the ε-algorithm has a wide range of applicability for this class of function.

If f(x) has midinterval singularities, and again we choose a typical example, with f(x) given by

$$f(x) = (x-s)^{\alpha}\, g(x), \text{ with } g(x) \text{ analytic in } [0,1] \text{ and } 0<s<1,$$

then $Q^{(m)}f$ has the form given by (Lyness and Ninham, 1967)

$$Q^{(m)}f = If + a_1h + a_2h^2 + \dots + a_ph^p + O(h^{p+1}) \qquad (16)$$

$$+ b_1^{(m)}\, h^{\alpha+1} + b_2^{(m)}\, h^{\alpha+2} + \dots + b_p^{(m)}\, h^{\alpha+p} + O(h^{p+\alpha+1}).$$

Here we notice that the coefficients $b_i^{(m)}$ are not independent of m. However, they have the property

$$b_\ell^{(m)} = y_\ell(ms) \ , \quad \text{where } y(ms+1) = y(ms).$$

If s is a rational number, then it will have a periodic binary expansion. Assume s is rational and let the length of the period be an integer $k_o$. By choosing the m sequence to be 1,2,4,... we will find that the $b_i^{(m)}$ have the desired periodic behaviour given by

$$b_i^{(2^j)} = b_i^{(2^{(j+k_o)})} .$$

Then, by defining $S_j$ as we have done previously, apart from the order terms we have

$$S_j = If + \sum_{i=1}^{p} a_i \beta_i^j + \sum_{i=1}^{p} c_i^{(j)} \delta_i^j , \qquad (17)$$

with $\beta_i = 2^{-i}$ , $c_i^{(j)} = b_i^{(2^j)}$ and $\delta_i = 2^{-(\alpha+i)}$ .

Comparing the form for $S_j$ given by equation (17) with that in equations (4) and (6) we see that ε-algorithm extrapolation will find If, but we also can see that we are likely to need a large number of terms in the sequence $\{S_j\}$, particularly when $k_o$ is large, and extrapolation will be difficult indeed when $k_o$ is irrational. We can conclude from this analysis that we really must know something about the location of s in order to decide whether or not to even try ε-extrapolation, but if we do have this information it is probably best in most cases to subtract out the singularity or transform it to the middle of the integration region.

## 3. Examples

We give a few actual numerical examples to illustrate some of the previous results. A more detailed comparison of various techniques for the forms given by equations (12) and (14) can be found in the paper by Chisholm, Genz and Rowlands (1972). The first example we consider is the integral

$$If = \int_0^1 x^{.75} \cos(x)\, dx = .44516493.$$

The ε-array obtained using a 2-point Gaussian-Legendre quadrature rule is given by (omitting the odd columns),

| | | |
|---|---|---|
| 0.447... | | |
| 0.4459... | 0.44501... | |
| 0.4454... | 0.445158... | 0.44516491... |
| 0.4452... | 0.445164... | |
| 0.44518... | | |

The ε-extrapolation is quite effective. The second example is an example of a principal value integral given by

$$If = \int_0^1 (x^2-0.01)^{-1}\,dx = -1.0033534.$$

Here the ε-array given below is obtained using an eight point Gauss-Legendre quadrature rule.

| | | | |
|---|---|---|---|
| 331.0... | | | |
| 15.252... | 86.667... | | |
| 107.535... | 54.483... | 155.75... | |
| -17.25... | -371.49... | -1.0033535... | -1.0033534... |
| -109.5... | -56.49... | -1.0033534... | |
| 15.2535... | 369.4... | | |
| 107.5342... | | | |

Notice the expected periodic behaviour exhibited by the first column of the displayed array and the dramatic convergence in column 3. The third example has an integrand with endpoint singularities as well as a midinterval one:

$$If = \int_0^1 (\ln x)(\ln|x-0.3|)(1-x)^{-0.56}\,dx = 2.3587825.$$

Again, the eight point Gauss-Legendre rule was used and the following array obtained,

| | | | |
|---|---|---|---|
| 2.19... | | | |
| 2.52... | 2.405... | | |
| 2.33... | 2.384... | 2.335... | |
| 2.39... | 2.373... | 2.3591... | 2.35870... |
| 2.35... | 2.365... | 2.358676... | 2.358745... |
| 2.37... | 2.362... | 2.358756... | |
| 2.358... | 2.360... | | |
| 2.361... | | | |

The extrapolation is fairly good as was expected.

For the multidimensional case the generalisation of equations (12), (14) and (16) has not been completely developed. Only the case for analytic functions (12) has been completely generalised (Lyness, 1965), but a limited number of examples and experimental results (Chisholm, Genz and Rowlands, 1972) indicate that the ε-algorithm should work well for the other cases as well, subject to restrictions already imposed. Some adaptive quadrature techniques to be used in conjunction with the ε-algorithm extrapolation have also been developed for one-dimensional and multidimensional cases (Genz, 1972).

## 4. The Infinite interval quadrature problem

There does not seem to be any similar analysis for the error expansions obtained for various sequences of approximations to integrals taken over infinite intervals which indicates that the ε-algorithm will be particularly effective. However, because of the powerful extrapolation properties of the ε-algorithm it is usually worthwhile applying it to any convergent sequence when no other method is available. Some limited success has been obtained using this method (Chisholm, Genz and Rowlands, 1972). There is also a simple generalisation of the ε-algorithm when the terms are written out in the Padé approximant determinant form (Shanks, 1955) which Gray, Atchison and McWilliams (1971) have developed and call the G-transformations. Although there is no longer a convenient algorithm like the $\varepsilon$-algorithm for computing the determinantal forms, the method seems to be a very effective one for the extrapolation of sequences of quadrature approximations in the infinite interval case.

## 5. Conclusion

We have considered the application of the $\varepsilon$-algorithm to specific sequences of quadrature approximations and found that it is frequently a very appropriate method to use, and have also delineated the cases where it will not be a particularly effective method.

## Acknowledgment

I would like to thank Professor James N. Lyness for his helpful suggestions about notation and order of presentation.

## References

Chisholm, J. S. R., Genz, A. and Rowlands, G. E. (1972). J. Comp. Phys. 10.

Davis, P. J. and Rabinowitz, P. (1967). "Numerical Integration", Blaidell, Waltham, Massachusetts.

Fox, L. (1967). Comp. Journal 10, 87-93.

Genz, A. (1972). To appear in Computer Physics Communications, 1972.

Gray, H. L., Atchison, T. A. and McWilliams, G. V. (1971). SIAM J. Num. Analysis 8, 365-381.

Joyce, D. C. (1971). SIAM Review 13, 435-490.

Lyness, J. N. (1965). Math. Comp. 19, 260-276.

Lyness, J. N. and Ninham, B. W. (1967). Math. Comp. 21, 162-178.

Romberg, W. (1955). Norske Vid. Selsk. Forh. 28, 30-36.

Shanks, D. (1955). J. Math. and Phys. 34, 1-42.

Wynn, P. (1956). Math. Comp. 10, 91-96.

Wynn, P. (1966). SIAM J. Num. Analysis 3, 91-122.

ON HADAMARD'S THEORY OF POLAR SINGULARITIES

W. B. Gragg

(Mathematics Dept., University of California, San Diego, U.S.A.)

Hadamard's theorem and consequences. Let $f(z) = \Sigma_0^\infty c_m z^m$ be a complex power series with radius of convergence $\rho_1 > 0$. Thus $\rho_1 = 1/\ell_1$ with $\ell_1 = \overline{\lim} |c_m|^{1/m}$. More generally let $c_{m,n} = \det (c_{m+i-j})$ $(i, j = 1, 2, \ldots, n)$, $\ell_n = \overline{\lim}_m |c_{m,n}|^{1/m}$ $(\ell_{-1} = 0$, $\ell_0 = 1)$ and $\rho_n = \ell_{n-1}/\ell_n$ $(= \infty$ if $\ell_n = 0)$. Let $r_{m,n}(z) = p_{m,n}(z)/q_{m,n}(z)$ be the Padé fraction of type $(m, n)$ for f. Thus $\deg p_{m,n} \leq m$, $\deg q_{m,n} \leq n$, g.c.d. $(p_{m,n}, q_{m,n}) = q_{m,n}(0) = 1$ and the Maclaurin expansion of $r_{m,n}$ agrees with f as far as possible. Then:

1. $$0 = \rho_0 < \rho_1 \leq \rho_2 \leq \rho_3 \leq \ldots \leq \rho_n \to \rho,$$

   the radius of meromorphy of f.

2. If $\{n(k)\}$, $0 = n(0) < n(1) < n(2) < \ldots$, is the set of indices n with $\rho_n < \rho_{n+1}$ then (the analytic continuation of) f has $\nu(k) = n(k) - n(k-1)$ poles $\pi_{n(k-1)+j}$, $j = 1, 2, \ldots, \nu(k)$, counting multiplicities, on the circle $|z| = \rho_{n(k)}$.

3. If $n \in \{n(k)\}$ then:

   a. $c_{m,n}/c_{m+1,n} \to_m \pi_1\pi_2 \ldots \pi_n$ and

   $$\overline{\lim}_m |c_{m,n}/c_{m+1,n} - \pi_1\pi_2 \ldots \pi_n|^{1/m} \leq \rho_n/\rho_{n+1};$$

b. $q_{m,n}(z) \to_m q_n(z) = \Pi_1^n (1 - z/\pi_k)$ uniformly on bounded sets and

$$\overline{\lim}_m |q_{m,n}(z) - q_n(z)|^{1/m} \le \rho_n/\rho_{n+1};$$

c. $r_{m,n}(z) \to_m f(z)$ uniformly on compact subsets of $D_{n+1} = \{z : |z| < \rho_{n+1}\} - \{\pi_1, \pi_2, \ldots, \pi_n\}$ and

$$\overline{\lim}_m |r_{m,n}(z) - f(z)|^{1/m} \le |z|/\rho_{n+1}, \quad z \in D_{n+1}.$$

Part 1 may be stated more precisely as $\ell_{n-1}\ell_{n+1} \le \ell_n$, $n \ge 0$; equivalently the sequence $\{\log \ell_n\}$ is concave. It implies that $\rho = 1/\ell$ with $\ell = \overline{\lim}\, \ell_n^{1/n}$. Parts 1 and 2 contain the essence of the Hadamard theory: beginning with the coefficients $\{c_m\}$ they allow one to deduce the meromorphic character of $f$. Their proof contains those of 3b (3a will follow on putting $z = 0$) and the converse assertions which assume from the outset that $f$ is meromorphic for $|z| < \rho$. The connection, in particular 3c, between the Hadamard theory and the Padé table is due to Montessus de Ballore.

Proof of the theorem. More generally let $\ell_n(z) = \overline{\lim}_m |v_{m,n}(z)|^{1/m}$ with 'unnormalized Padé denominators'

$$v_{m,n}(z) = d \begin{pmatrix} 1 & z & \cdots & z^n \\ c_{m+1} & c_m & \cdots & c_{m-n+1} \\ \vdots & \vdots & & \vdots \\ c_{m+n} & c_{m+n-1} & \cdots & c_m \end{pmatrix}$$

$$= c_{m,n} + \dots + (-1)^n c_{m+1,n} z^n.$$

Thus $\ell_n = \ell_n(0)$. If $c_{m,n} \neq 0$ then from the elementary theory of the Padé table $v_{m,n}(z) = c_{m,n}q_{m,n}(z)$ and $r_{m,n}(z) = f(z) + O(z^{m+n+1})$. Moreover if $c_{m,n}c_{m+1,n} \neq 0$ then the following Frobenius identities are valid:

$$q_{m+1,n}(z) - q_{m,n}(z) = z\,\frac{c_{m+1,n+1}v_{m,n-1}(z)}{c_{m,n}c_{m+1,n}} \qquad (1)$$

$$r_{m+1,n}(z) - r_{m,n}(z) = (-1)^n \frac{c_{m+1,n+1}}{c_{m,n}}\,\frac{z^{m+n+1}}{q_{m,n}(z)q_{m+1,n}(z)}. \qquad (2)$$

Now $\ell_2 = \overline{\lim}\,|c_{m-1}c_{m+1} - c_m^2|^{1/m}$ and by a simple estimation $\ell_n(z) \le \ell_1^n$. A 'delicate' point of the proof is <u>Hadamard's lemma</u>: if $\ell_{n-1}\ell_{n+1} < \ell_n^2$ then $\ell_n = \lim_m |c_{m,n}|^{1/m}$ is a regular limit. Pólya has given an elegant proof for the crucial case $n = 1$ $(\ell_2 < \ell_1^2)$. The general case uses the identity $c_{m-1,n}c_{m+1,n} + c_{m,n-1}c_{m,n+1} = c_{m,n}^2$ which results on comparing coefficients of $z^n$ in (1). Thus $\overline{\lim}_m\,|c_{m-1,n}c_{m+1,n} - c_{m,n}^2|^{1/m} = \overline{\lim}_m\,|c_{m,n-1}c_{m,n+1}|^{1/m} \le \ell_{n-1}\ell_n < \ell_n^2$ and one applies the previous result to the sequence $\{c_{m,n}\}_{m=0}^{\infty}$.

The theorem is proved by induction on $k$. Suppose $f$ has been continued to the disk $|z| < \rho_{n(k)+1}$. Let $q_{n(s)}(z) = 1 + b_1^{(s)}z + \dots + b_{n(s)}^{(s)}z^{n(s)}$ and $f^{(s)}(z) = q_{n(s)}(z)f(z) = \Sigma_{m=0}^{\infty} c_m^{(s)}z^m$, $s = 0, 1, \dots, k$. Then $f^{(s)}$ is holomorphic for $|z| < \rho_{n(s)+1}$ so $\overline{\lim}_m\,|c_m^{(s)}|^{1/m} \le 1/\rho_{n(s)+1}$. One now uses

$c_m^{(s)} = c_m + b_1^{(s)} c_{m-1} + \dots + b_{n(s)}^{(s)} c_{m-n(s)}$ and column operations to replace the elements $c_{m+i-j}$ in the determinant for $v_{m,n(k)+n}(z)$ by elements $c_{m+i-j}^{(s)}$ with 'slowest growth' as $m \to \infty$. There results a matrix whose first $n + 1$ columns contain $c_{\cdot}^{(k)}$, the next $\nu(k)$ columns contain $c_{\cdot}^{(k-1)}$, and so on until finally the last $\nu(1)$ columns contain $c_{\cdot}^{(0)} = c_{\cdot}$. Expansion along the top row and use of the general Laplace expansion together with $\rho_{n(1)} < \rho_{n(2)} < \dots < \rho_{n(k)}$ then provide the estimate

$$\ell_{n(k)+n}(z) \le 1/\rho_{n(1)}^{\nu(1)} \cdots \rho_{n(k)}^{\nu(k)} \rho_{n(k)+1}^{n}, \quad n \ge 0, \tag{3}$$

which is classical for $z = 0$ and reduces to $\ell_n(z) \le \ell_1^n$ for $k = 0$.

Let $z = 0$. Suppose equality in (3) for $n = 0, 1, \dots, \nu(k+1)$, but strict inequality for $n = \nu(k+1) + 1$. That is $\rho_{n(k)} < \rho_{n(k)+1} = \dots = \rho_{n(k+1)} < \rho_{n(k+1)+1}$. In particular

$$\ell_{n(k+1)-1}(z)\ell_{n(k+1)+1} < \ell_{n(k+1)}^2 \tag{4}$$

for all $z$. By Hadamard's lemma $\ell_{n(k+1)}$ is a regular limit and $c_{m,n(k+1)} \ne 0$ for large $m$. Assertion 3 for $n = n(k + 1)$ (with perhaps different values $\pi_1', \pi_2', \dots, \pi_{n(k)}'$) now follows from (4) by majorization in (1) and (2).

Some more refined estimates will now be developed to show that $q_{n(k)}$ divides $q_{n(k+1)}$ (whence the primes may be deleted) and that $f^{(k+1)}(z) = q_{n(k+1)}(z)f(z)$ is holomorphic for

$|z| < \rho_{n(k+1)+1}$. For $n = \nu(k+1)-1$ ($\geq 0$) expansion along the top row of the modified determinant and estimation gives

$$v_{m,n(k+1)-1}(z) = \sum_{j=1}^{k+1} z^{\sigma(j+1,k+1)} v^{(k)}_{m,\nu(j)-1}(z) q_{n(j-1)}(z)$$

with $\sigma(j, k) = \Sigma_j^k \nu(i)$, $\deg v^{(k)}_{m,\nu(j)-1} \leq \nu(j)-1$ and

$$\overline{\lim}_m |v^{(k)}_{m,\nu(j)-1}(z)|^{1/m} \leq \rho_{n(j)} \ell_{n(k+1)}. \tag{5}$$

Likewise, with $n = \nu(k + 1)$ and division by $c_{m,n(k+1)}$,

$$q_{m,n(k+1)}(z) = \sum_{j+1}^{k+1} z^{\sigma(j+1,k+1)} q^{(k)}_{m,\nu(j)}(z) q_{n(j-1)}(z) \tag{6}$$

with $\deg q^{(k)}_{m,\nu(j)} \leq \nu(j)$, $q^{(k)}_{m,\nu(j)}(0) = \delta_{j,k+1}$ and

$$\overline{\lim}_m |q^{(k)}_{m,\nu(j)}(z)|^{1/m} \leq \rho_{n(j)}/\rho_{n(k+1)} \quad (< 1,\ j \leq k). \tag{7}$$

Comparison with (1) now indicates, and it follows by a division algorithm, that

$$q^{(k)}_{m+1,\nu(j)}(z) - q^{(k)}_{m,\nu(j)}(z) = z \frac{c_{m+1,n(k+1)+1} v^{(k)}_{m,\nu(j)-1}(z)}{c_{m,n(k+1)} c_{m+1,n(k+1)}} \tag{8}$$

From (5) and (8), by majorization, $q^{(k)}_{\nu(j)}(z) = \lim_m q^{(k)}_{m,\nu(j)}(z)$ exists and

$$\overline{\lim}_m |q^{(k)}_{m,\nu}(z) - q^{(k)}_{\nu(j)}(z)|^{1/m} \leq \rho_{n(j)}/\rho_{n(k+1)+1}. \tag{9}$$

Then from (6) and (7) one finds $q^{(k)}_{\nu(j)} = \delta_{j,k+1} q_{n(k+1)}/q_{n(k)}$, proving the divisibility. (Hence <u>a posteriori</u>

$$\overline{\lim}_m \, |q^{(k)}_{m,\nu(j)}(z)|^{1/m} \le \rho_{n(j)}/\rho_{n(k+1)+1}$$

for $j = 0, 1, \ldots, k$.) Now (6) and the 'Padé property' provide

$$f^{(k+1)}(z) = [q^{(k)}_{\nu(k+1)}(z) - q^{(k)}_{m,\nu(k+1)}(z)]f^{(k)}(z)$$

$$- \sum_{j=1}^{k} z^{\sigma(j+1,k+1)} q^{(k)}_{m,\nu(j)}(z) f^{(j-1)}(z)$$

$$+ p_{m,n(k+1)}(z) + O(z^{m+n(k+1)+1}).$$

Finally, since $n(k + 1) \ge 1$ for $k \ge 0$, one may equate coefficients of $z^{m+1}$ and apply the estimates (9) and $\overline{\lim}_m \, |c^{(j)}_m|^{1/m} \le 1/\rho_{n(j)+1} = 1/\rho_{n(j+1)}$ $(j = 0, 1, \ldots, k)$ to deduce that $\overline{\lim}_m \, |c^{(k+1)}_m|^{1/m} \le 1/\rho_{n(k+1)+1}$.

To complete the proof suppose that $f$ has <u>fewer</u> than $n(k+1)$ poles in $|z| < \rho_{n(k+1)+1}$. Then there is a polynomial $p_n(z)$, of degree $n < \nu(k + 1)$, such that $g(z) = p_n(z)f^{(k)}(z) = p_n(z)q_{n(k)}(z)f(z)$ is holomorphic for $|z| < \rho_{n(k+1)+1}$. As in the argument used to establish (3), but now with <u>one more reduction</u> to introduce $c^{(k+1)}_{\cdot}$, $\ell_{n(k)+n+1} \le 1/\rho^{\nu(1)}_{n(1)} \cdots \rho^{\nu(k)}_{n(k)}\rho^{n}_{n(k+1)}\rho_{n(k+1)+1} < 1/\rho^{\nu(1)}_{n(1)} \cdots \rho^{\nu(k)}_{n(k)}\rho^{n+1}_{n(k+1)}$. This strict inequality contradicts the definition of $\nu(k + 1)$. Thus $f$ has exactly $n(k + 1)$ poles in $|z| < \rho_{n(k+1)+1}$ and, since $f^{(k+1)}(z) = q_{n(k+1)}(z)f(z)$ is bounded at these points they must be the zeros of $q_{n(k+1)}$. Now $|\pi_{n(k)+j}| \geq \rho_{n(k)+1} = \rho_{n(k+1)}$ and from 3a, $|\pi_{n(k)+1}\pi_{n(k)+2} \cdots \pi_{n(k+1)}| = \ell_{n(k)}/\ell_{n(k+1)} = \rho_{n(k+1)}^{\nu(k+1)}$. Consequently $|\pi_{n(k)+j}| = \rho_{n(k+1)}$ for $j = 1, 2, \ldots, \nu(k+1)$.

References

Gragg W. B., (1972). SIAM Review, 14, 1.

Hadamard J., (1892). J. Math. Pures. Appl., 8, 101.

de Montessus de Ballore R., (1902). Bull. Soc. Math. France, 30, 28.

Pólya G., (1925). Enseignement Math., 24, 76.

# TRUNCATION ERROR BOUNDS FOR CONTINUED FRACTIONS AND PADÉ APPROXIMANTS

William B. Jones

(University of Colorado, Boulder, Colorado, U.S.A.)

In many problems it is important to have sharp estimates of the truncation error $|f - f_n|$ in the approximation of a convergent continued fraction

$$(1) \qquad f = K(a_n/b_n) = \frac{a_1}{b_1} + \frac{a_2}{b_2} + \frac{a_3}{b_3} + \cdots$$

by its nth approximant $f_n$. Here the elements $a_n, b_n$ denote either complex constants or functions $a_n(z), b_n(z)$ of a complex variable $z$. There are two main classes of error bounds: (a) a priori bounds are expressed directly in terms of the $a_n, b_n$ or in terms of parameters associated with these elements. For results of this type see (Field and Jones, 1972) and references contained therein. (b) a posteriori bounds, usually of the form

$$(2) \qquad |f - f_n| \leq M_n |f_n - f_{n-1}| ,$$

are determined after the approximants $f_{n-1}, f_n$ have been calculated. This paper provides a brief survey of truncation error bounds of a posteriori type. The relation between these results and the Padé table is quite strong. In a normal Padé table the approximants of the corresponding continued fraction of the form

$$(3) \qquad f(z) = \frac{a_0}{1} + \frac{a_1 z}{1} + \frac{a_2 z}{1} \cdots \text{ (complex } a_n \neq 0)$$

fill the stairlike sequence of squares [0,0],[1,0],[1,1],[2,1], [2,2],... (Wall, 1948, Theorem 96.1). For the Padé table of an arbitrary power series $f(z)$, Magnus (1962a,b) has introduced the P-fractions with the property that $z^{-s}$ times each approximant lies in the diagonal sequence $[n, n-s]$ and, conversely, every entry in the Padé table of $f(z)$ is obtained in this manner.

A unifying property of all results given below is the following: A sequence of complex numbers $\{f_n\}$ is called a simple sequence if

$$(4) \qquad |f_{n+m} - f_n| \leq C|f_n - f_{n-1}|, \quad m \geq 0, \ n \geq 2 ,$$

for some constant $C > 0$, independent of $n$. Simple sequences are a natural generalization of real number sequences with the nesting property

$$(5) \qquad f_{2n-2} \leq f_{2n} \leq f_{2n+1} \leq f_{2n-1} .$$

A simple sequence may not converge (e.g., $f_n = (-1)^n$) and, moreover, a convergent sequence may not be simple (e.g., $f_n = 1/n$; here the convergence is too slow for the sequence to be simple). When such a constant $C$ can be found, the right side of (4) is an a posteriori error bound, provided $f = \lim f_n$ exists. Table 1 contains constants $C$ corresponding to most of the known cases. Proofs and further discussion may be found in references below. Letters in the left column of Table 1 will be used to identify that particular result.

Perhaps the first known examples of continued fractions with simple sequences of approximants are those of the form $K(a_n/1)$ with $a_n > 0$ and $K(1/b_n)$ with $b_n > 0$; in such cases the approximants are positive real numbers satisfying (5) with simple se-

Table 1. Simple Sequences $\{f_n\}$ from Continued Fractions

| | $f_n$ = nth Approximant of: | C = Simple Sequence Constant |
|---|---|---|
| $B_1$ | $\frac{a_1}{1} + \frac{a_2}{1} + \cdots, \quad \lvert a_n\rvert \le \frac{1}{4} - \varepsilon, \quad 0 < \varepsilon < 1/4$ | $C = (1/2\varepsilon) - 1$ |
| $B_2$ | $\frac{1}{b_1} + \frac{1}{b_2} + \cdots, \quad \lvert b_n\rvert \ge 2 + \varepsilon, \quad \varepsilon > 0$ | $C = d/(1-d), \quad d = 1 + \frac{\varepsilon}{2} - [(1 + \frac{\varepsilon}{2})^2 - 1]^{1/2}$ |
| M | $\frac{a_1}{b_1} + \frac{a_2}{b_2} + \cdots, \quad \left\lvert \frac{a_n}{b_n b_{n-1}} \right\rvert \le r(1-r), \ 0 < r < 1/2$ | $C = r/(1-2r)$ |
| HP | $\frac{a_1}{z} + \frac{a_2}{1} + \frac{a_3}{z} + \frac{a_4}{1} + \cdots, \quad a_n > 0, \quad \lvert\arg z\rvert < \pi$<br>(S-fraction) | $C(z) = \begin{cases} 1, & \lvert\arg z\rvert \le \pi/2 \\ \tan[(1/2)\arg z], & \pi/2 < \lvert\arg z\rvert < \pi \end{cases}$ |
| J | $1 + d_o z + \frac{z}{1 + d_1 z} + \frac{z}{1 + d_2 z} + \cdots, \quad d_n > 0,$<br>(T-fraction) $\quad \lvert\arg z\rvert < \pi$ | $C(z) = \begin{cases} 1, & \lvert\arg z\rvert \le \pi/2 \\ \sec[(1/2)\arg z], & \pi/2 < \lvert\arg z\rvert < \pi \end{cases}$ |
| JT | $\frac{a_1}{b_1} + \frac{a_2}{b_2} + \cdots, \quad \gamma_o, \ \theta \text{ real}, \ 0 < \lvert\theta\rvert < \pi$<br>$\gamma_n = \arg a_n - \gamma_{n-1} - \theta \pmod{2\pi}$<br>$b_n e^{-i\gamma_n} \in \begin{cases} D[0,\theta], & 0 < \theta < \pi \\ D[\theta,0], & -\pi < \theta < 0 \end{cases}$<br>where<br>$D[\alpha,\beta] \equiv \{z: z = 0 \text{ or } \alpha \le \arg z \le \beta\}$ . | $C(\theta) = \begin{cases} 1, & 0 < \lvert\theta\rvert < \pi/2 \\ \sec(\lvert\theta\rvert - \pi/2), & \pi/2 < \lvert\theta\rvert < \pi \end{cases}$ |

quence constant $C = 1$. These results are obtainable as special cases of HP in Table 1, with $z$ real and positive. The first examples with complex elements $a_n, b_n$ were obtained by Blanch (1964), using comparison relations for continued fractions ($B_1$ and $B_2$ in Table 1). An improvement of these results was obtained by Merkes (1966) from an analysis based on chain sequences (M in Table 1). In the same year Henrici and Pfluger (1966) developed error bounds for S-fractions (or series of Stieltjes) by studying best inclusion regions (HP in Table 1). Using a similar argument, Jefferson (1969) obtained inclusion regions and error bounds for T-fractions (J in Table 1). A generalization of both HP and T was obtained by JT (Jones and Thron, 1971). In additon to HP and J, the result given by JT can also be applied to continued fractions of Gauss, to large classes of J-fractions, to continued fractions of the form $K(1/b_n)$ where $|\arg b_n| \le (\pi/2) - \varepsilon$, $\varepsilon > 0$ or $b_n = 0$, and to other examples. The conditions on the $a_n, b_n$ in JT are invariant in form under equivalence transformation of continued fractions.

When an S-fraction (or series of Stieltjes) is a holomorphic function $f(z)$ at the origin, one can obtain sharper error bounds than those of HP. Common (1968) obtained such bounds for positive and negative real values of $z$, and Baker extended this to complex $z$ by considering best inclusion regions. However, Baker's analysis did not provide a simple method to calculate the error bound, since it did not show that the lens-shaped inclusion region is convex (a property that greatly facilitates calculating the diameter of the region). This was done in an independent study by Gragg (1968) for the case with f(z) holomorphic in the unit disk $\{z: |z| < 1\}$. In a second paper Gragg (1970) extended his analysis to include functions f(z) holomorphic in the complex plane cut along an arbitrary finite interval of the real axis.

## References


Baker, George A., Jr. (1969). J. Mathematical Physics 10, pp 814 - 820.

Blanch, G. (1964). SIAM Rev. 6, pp 383 - 421.

Common, A. K. (1968). J. Mathematical Physics 9, pp 32 - 38.

Field, David A. and William B. Jones (1972). Numer. Math. To appear.

Gragg, W. B. (1968). Numer. Math. 11, pp 370 - 379.

______. (1970). Amer. Math. Soc. Bull. 76, pp 1091 - 1094.

Henrici, P. and P. Pfluger (1966). Numer. Math. 9, pp 120 - 138.

Jefferson, T. H. (1969). SIAM J. Numer. Anal. 6, pp 359 - 364.

Jones, William B. and W. J. Thron (1971). SIAM J. Numer. Anal. 8, pp 693 - 705.

Magnus, A. (1962 a). Math. Zeitschr. 78, pp 361 - 374.

______. (1972 b). Math. Zeitschr. 80, pp 209 - 216.

Merkes, E. P. (1966). SIAM J. Numer. Anal. 3, pp 484 - 496.

Wall, H. S. (1948). "Analytic Theory of Continued Fractions", Van Nostrand, Princeton.

# USE OF THE PADE TABLE FOR APPROXIMATE LAPLACE TRANSFORM INVERSION

I. M. Longman

(Department of Environmental Sciences, Tel-Aviv University, Israel)

## 1. Introduction

It is well known that rational function approximations to a function $\bar{f}(p)$ of the Laplace transform variable p can often (by inversion) be used to obtain useful approximations to the inverse f(t). As examples of this approach reference may be made to the work of Luke (1962, 1964, 1969), Longman (1966, 1967, 1968, 1970), Akin and Counts (1969a,b), Murphy (1971). In many cases the Padé table provides a convenient means of obtaining rational approximations suitable for this purpose. This is especially so when the Padé table is normal, since for this case it can be readily computed by a recursive method developed by the author (Longman, 1971).

For most physical problems where the solution f(t) does not tend to infinity with t, it is desirable that our rational approximations to $\bar{f}(p)$ have no poles in the right-hand half of the p-plane $Re(p) > 0$, so that we have no terms tending to infinity after inversion. Unfortunately, except in the case where we start from a series of Stieltjes, or an equivalent series, our knowledge of the expected locations of poles of the Padé approximants is limited. However in two physical problems studied by the author most of the leading diagonal approximants considered had this property.

Owing to severe space limitations, a full discussion cannot be given here. It is merely possible to give brief reference to two examples of important physical problems treated by inversion of Padé approximants. The details are being published elsewhere.

## 2. A Problem in Viscoelasticity

In this problem (Longman, 1972a) we are led to consider the inversion of

$$\bar{f}(p) = (1/p)\,\exp\left[-p(1+\sigma p)^{-\frac{1}{2}}\right] \tag{1}$$

for various values of the parameter $\sigma$.

The method adopted was to expand $p\bar{f}(p)$ in a Maclaurin series and generate from this the Padé table by use of a recursive scheme due to the author (Longman, 1971). Successive leading diagonal elements were divided by p and inverted, and the results tabulated in columns for a range of values of t for each $\sigma$ considered. Comparison of the columns afforded a measure of the accuracy achieved. In most cases the poles were found to lie in the left-hand half-plane $\mathrm{Re}(p) \leq 0$. However, for a few values of $\sigma$ certain columns contained results significantly different from their neighbours, and this was found to be due to the presence of poles in the right-hand half-plane $\mathrm{Re}(p) > 0$.

A special difficulty here was that for the case $\sigma=1$ the Padé table is not normal, and the method for its computation due to the author was not applicable.

## 3. A Problem in Electrical Network Theory

For details the reader must again be referred to a fuller account (Longman, 1972b). The function here to be inverted had the form

$$\bar{f}(p) = 1/\left[pF(p)\right] \tag{2}$$

where

$$F(p) = \int_0^\lambda e^{-u^2/4}\cosh(up^{\frac{1}{2}})\,du \Big/ \int_0^\lambda e^{-u^2/4}\,du\ , \tag{3}$$

$\lambda$ being a positive parameter, and the inversion was carried out by use of successive leading diagonal elements of the Padé table obtained from the Maclaurin expansion of F(p). Again in most cases poles of the approximations thus obtained for $\bar{f}(p)$ were in the left-hand half-plane $Re(p) \leqslant 0$.

## Acknowledgment

The computations for the work described here were carried out at the Computation Centre of Tel-Aviv University.

## References

Akin J. E. and Counts J., (1969a). SIAM J. App. Math. 17, 1035.

Akin J. E. and Counts J., (1969b). J. App. Mech. 36, 420.

Longman I. M., (1966). Bull. Seism. Soc. Am. 56, 1045.

Longman I. M., (1967). Geophys. J. Roy. Astr. Soc. 13, 103.

Longman I. M., (1968). J. Inst. Maths. Applics. 4, 320.

Longman I. M., (1970). Geophys. J. Roy. Astr. Soc. 21, 295.

Longman I. M., (1971). Internat. J. Computer Math. B3, 53.

Longman I. M., (1972a). J. Comp. Phys., in press.

Longman I. M., (1972b). SIAM J. App. Math., in press.

Luke Y. L., (1962). Proc. 4th U.S. Nat. Cong. App. Mech., 269.

Luke Y. L., (1964). Quart. J. Mech. App. Math. 17, 91.

Luke Y. L., (1969). In "The Special Functions and their Approximations", pp.255-268, Academic Press N.Y.

Murphy J. A., (1971). J. Inst. Maths. Applics. 7, 138.

# PADE APPROXIMANT AND HOMOGRAPHIC TRANSFORMATION OF RICCATI'S PHASE EQUATIONS

A. Ronveaux
(Facultés Universitaires de Namur, Namur, Belgium)

## 1. Introduction

A well known technique to obtain approximate solutions of the Schrödinger equation with a potential V(r) consists of transforming the radial equation into a linear integral equation. However, for a large class of quantum mechanical problems, the knowledge of the wave functions is not essential. The wave function is only used to compute some functionals of the interaction: phase shift, scattering length, spectrum etc.

There exists another technique which replaces the Schrödinger equation by an appropriate equation giving directly the functional of the potential, bypassing the computation of the wave function. This not widespread technique, i.e. replacing a second order linear differential equation by a first order non linear equation, gives the so called "phase equation".

## 2. Phase equation

The solution of the differential equation (Schrödinger type):

$$u'' + qu = Vu \tag{1}$$

can be written

$$u(r) = c_1(r)u_1(r) + c_2(r)u_2(r), \tag{2}$$

where $u_i(r)$ (i=1,2) are linearly independent solutions of the unperturbed equation ($V(r) \equiv o$) with Wronskian
$W = u_1u'_2 - u_2u'_1$ .

The ratio $S = \frac{c_2}{c_1}$ called the "phase", satisfies the Riccati equation

$$S' = \frac{V}{W}(u_1 + Su_2)^2 \quad . \tag{3}$$

In the same way, the solution of the differential system (Dirac type)

$$\vec{X} = |A|\vec{X} + |V|\vec{X} \quad , \quad \vec{X} = \begin{vmatrix} g \\ f \end{vmatrix} \quad , \quad |V| = V(r) \begin{vmatrix} o & -1 \\ 1 & o \end{vmatrix}, \tag{4}$$

can be written

$$\vec{X} = c_1\vec{u}_1 + c_2\vec{u}_2 \quad , \quad \vec{u}_i = \begin{vmatrix} u_i \\ \bar{u}_i \end{vmatrix} \quad , \tag{5}$$

where $\vec{u}_i (i=1,2)$ are solutions of the unperturbed system ($V \equiv o$) with $W = u_1\bar{u}_2 - u_2\bar{u}_1 \neq o$.

The phase equation ($S = \frac{c_2}{c_1}$) is now (Ronveaux 1970)

$$S' = \frac{V}{W} \left| (u_1 + Su_2)^2 + (\bar{u}_1 + S\bar{u}_2)^2 \right| . \tag{6}$$

## 3. Review of some phase equations

Let us indicate, without derivation, some important phase equations obtained by identification of the phase S with appropriate functionals.

| | |
|---|---|
| Phase Shift $\delta_L$ | $t' = -k^{-1}V(\hat{j}_L - t\,\hat{n}_L)^2$ (7) |
| L=o (Morse and Allis 1933) | $t(o), t(\infty) = \tan \delta_L$ |
| L≠o (Kynch 1952) | $\hat{j}_L, \hat{n}_L$ Riccati-Bessel functions. |
| (Calogero 1963) | |
| (Drukarev 1965) | |
| Generalized Scattering | $a_L' = (2L+1)^{-1}V(r^{L+1} - a_L r^{-L})^2$ |
| length $a_L$ | $a(o)=o, a(\infty) = a_L$ (8) |
| L=o (Spruch 1962) | |
| L≠o (Calogero 1963) | |

| | |
|---|---|
| Relativistic scattering length $a^{(R)}$ | $a' = 2mh^{-2}V\|(r-a)^2 + h^2(2mcr)^{-2}a^2\|$ |
| $j=\frac{1}{2}$ (Ronveaux 1970) | $\begin{cases} a(o) = o \\ a(\infty) = a^{(R)} \end{cases}$ (9) |
| Spectrum of central potential | $S' = (2q)^{-1}V(e^{qr} - Se^{-qr})^2$ (10) |
| $E = -q^2$ $(q>o)$<br>L=o (Calogero 1963) | $S(o)=1$, $S(\infty)=\infty$ <=> bound state of energy E |
| Spectrum of spheroidal potential (prolate case) | $S' = f^2 g(1-S \coth^{-1}\xi)^2$ (11) |
| L=o, m=o (Ronveaux 1972) | $S(1)=o$, $S(\infty)=\infty$ <=> bound state, f=focal distance, $g(\xi)$ prolate potential. |

These phase equations yield two different mathematical problems.

(a) The approximate location of the pole of the solution characterizing properties of the functionals (condition on V to ensure bound states, phase shift smaller that $\frac{\pi}{2}$ ...)

(b) The approximation of the solution (bound on the phase shift, scattering length...). The connection with Padé approximant will be only shown for problem of type b).

## 4. Approximations on the phase (interaction $\lambda V$)

Two different procedures in relation with Padé can be applied.

(a) Picard's iteration in the S equation, generates a power series in $\lambda$ which is reminiscent of the Born series. Unfortunately the second iteration (improved Born) is sometimes worse than the first Born approximations (Calogero 1967). Conversion of this power series in a Padé series was recently considered (Kloet and Zimerman 1970) for the scattering length of a repulsive singular potential with an analyticity assumption on V(r).

(b) Homographic transformation (Ronveaux 1968) of the phase gives directly $|N,N|$ Padé approximant reminiscent of approximation obtained by different techniques: distorted wave (Swan 1960, Brysk 1962), quasiparticle method (Weinberg 1963), distortion operator (Michalik 1968), Fredholm approximants (Newton 1968).

## 5. Homographic transformations and Padé approximant

The homographic group is the group of the Riccati equation and the iteration series in any transformed Riccati equation will generate a meromorphic series in $\lambda$ for the initial phase. Among all these transformations, a particular one will however furnish exact bounds and not only approximations.

The homographic transformation

$$T = Su_1(u_1+Su_2)^{-1} \tag{12}$$

transforms eq(3) in

$$T' = -Wu_1^{-2}T^2 + W^{-1}\lambda Vu_1^2 . \tag{13}$$

The Wronskian W being a constant (let $W>o$), iterations in eq(13) give now a meromorphic S series from eq(12).

The T equation, as compared to any transformed equation, is remarkable for two reasons.

(a) The V terms are dissociated from the T function.

(b) The signs of each term are well defined and do not depend on the sign of T.

Even with this particular structure, the situation is not easy to analyse in general. Bounds on T and S can only be given when V(r) is positive (negative) definite and of short range. ($V(r) = o$ $r>R$). It is clear from eq(13) that the sign of V and the position of the zero's of $u_1$ are essential in the discussion. (Ronveaux 1968).

## 6. Conclusion

We are aware of the limitation of the technique (the difficulty to extend to three dimensions) but we think also that it is useful to explore this frontier. We feel that in some cases (see section 5) the homographic approach to Padé could determine the sign of the error in the convergence of some Padé approximants.

## REFERENCES

Brysk, H., (1962), Phys. Rev. 126, 1589

Calogero, F., (1967), "Variable phase approach to potential scattering", Academic Press, N.Y.

Kloet, W.M., and Zimerman, A.H., (1970), Il nuovo cimento, 69A, 564

Michalik, B., (1968), Ann. Phys. 48A, 197

Ronveaux, A., (1968), Compt. Rend. 266A, 172; 266A, 478; 266A, 306

Ronveaux, A., (1970), Nucl. Phys. 18, 245

Swan, P., (1960), Nucl. Phys. 18, 245

Weinberg, S., (1964), Phys. Rev. 133B, 232

# PROPERTIES OF $I_{MN}$ APPROXIMANTS

V. Zakian

(University of Manchester Inst. of Science and Technology, England)

An $I_{MN}$ approximant of a function $f(t)$, $0 \leq t < \infty$, is defined by the linear functional

$$I_{MN} f = \int_0^\infty f(\lambda) \sum_{i=1}^{N} \frac{K_i}{t} e^{-\lambda \alpha_i / t} \, d\lambda$$

where $\alpha_i$, $K_i$ are constants and $M<N$. A subclass of $I_{MN}$ approximants, called $I_{MN}^p$ approximants are subject to the condition that if $f(\lambda) = e^{-\lambda}$, then $I_{MN}^p f$ is the corresponding (M,N) Padé approximant, N being the degree of denominator. Although in general not rational functions, $I_{MN}^p$ approximants have some of the remarkable properties of Padé approximants, and have significant applications in numerical analysis.

It is easy to verify that

$$I_{MN} f = \sum_{i=1}^{N} \frac{K_i}{t} F(\alpha_i / t) \tag{1}$$

where $F(s)$ is the Laplace transform of f, that is

$$F(s) = \int_0^\infty f(\lambda) \, e^{-\lambda s} \, d\lambda \quad .$$

Evidently (1) provides a formula for approximating $f(t)$ from numerical values of the Laplace transform $F(s)$. Preliminary work (Zakian, 1971) shows that the formula is the basis of unconditionally stable recursions for solving differential equations.

Some of the principal properties of $I_{MN}$ approximants are stated in theorems below; the proofs are contained in a more complete report (Zakian, 1972).

Let $E_{MN}f = f(t) - I_{MN}f$ define the truncation error, and let

$$T_n = \{t : \frac{|t|}{t} \min_i\{\mathrm{Re}(\alpha_i)\} > \hat{\sigma}_n |t|\}$$

$$\hat{\sigma}_n = \inf\{\sigma : f^{(k)}(\lambda) = O(e^{\sigma\lambda}),\ \lambda\to\infty;\ k=0,1,2,\ldots,n\}$$

where $\sigma$ is a real variable.

For a given t, the operator $I_{MN}$ is defined by an infinite integral, therefore $I_{MN}$ exists if and only if the integral converges.

Theorem 1

Let f be continuous on $[0,\infty)$ and $f(\lambda) = O(e^{\sigma\lambda})$, $\lambda\to\infty$; then $I_{MN}$ exists for every $t \in T_o$.

We note that if $\min_i\{\mathrm{Re}(\alpha_i)\} > 0$, and $\sigma<0$, then $T_o$ contains all $t\geq 0$.

Theorem 2

With the hypothesis of Theorem 1, $I_{MN}f$ is analytic in $T_o - \{0\}$.

Although the operator $I_{MN}$ may not exist for values of t outside $T_o$, it may happen that the function $I_{MN}f$ is defined by analytic continuation for values of t not in $T_o$. In such circumstances $I_{MN}f$ is not necessarily an approximant to f(t). For example, if $f(t) = e^t$, then $E_{MN}f \to \infty$ as $t\to\infty$, and thus for large t, although $I_{MN}f$ is analytic it is not an approximant to f(t).

Theorem 3

For all $t\geq 0$,

$$I^p_{MN}\, t^k = t^k, \qquad k=0,1,2,\ldots,(M+N)$$

$$I^p_{MN}\, t^k = t^k \sum_{i=1}^{N} \frac{K_i\, k!}{\alpha_i^{k+1}} , \qquad k > (M+N)$$

if and only if $\min_i\{\mathrm{Re}(\alpha_i)\} > 0$ .

Theorem 4

With the hypotheses of Theorem 1, if $\sigma<0$, then

$$E^p_{MN} f = O(t^{-(N-M)}) , \quad t\to\infty .$$

Theorem 5

Let $n\leq(M+N)$, let $f, f^{(1)},\ldots,f^{(n+1)}$ be continuous on $[0,\infty)$ and let $f^{(k)}(\lambda) = O(e^{\sigma\lambda})$, $\lambda\to\infty$, for $k=0,1,2,\ldots,n$; then for all $t \in T_n$

$$I^p_{MN} f = \sum_{k=0}^{n} \frac{t^k}{k!} f^{(k)}(0) + R_{MN,n}(t)$$

$$R_{MN,n}(t) = t^{n+1} \int_0^\infty f^{(n+1)}(\lambda) \sum_{i=1}^{N} \frac{K_i}{t\, \alpha_i^{n+1}} e^{-\lambda\alpha_i/t}\, d\lambda .$$

Theorem 6

With the hypotheses of Theorem 5 and assuming that $f^{(n+1)}(\lambda)= O(e^{\sigma\lambda})$, $\lambda\to\infty$; if $\min_i\{\mathrm{Re}(\alpha_i)\} > 0$ then

$$R_{MN,n}(t) = O(t^{n+1}) , \quad t\to 0 .$$

Theorem 7

With the hypotheses of Theorem 5, for every $t \in T_n \cap [0,\infty)$

$$E^p_{MN} f = \int_0^\infty f^{(n+1)}(\lambda)\, K_{MN,n}(\lambda,t)\, d\lambda$$

$$K_{MN,n}(\lambda,t) = \frac{(t-\lambda)^n}{n!}\, \Pi(\lambda,t) - \sum_{i=1}^{N} \frac{K_i}{\alpha_i^{n+1}} t^n e^{-\lambda\alpha_i/t}$$

$$\Pi(\lambda,t) = 1 \; , \quad 0 \leq \lambda \leq t$$
$$= 0 \; , \quad t < \lambda < \infty$$

Theorem 8

Let $n \leq M+N$, let $f^{(k)}(\lambda)$ be continuous on $[0,\infty)$ and $f^{(k)}(\lambda) = O(e^{\sigma\lambda})$, $\lambda \to \infty$, for $k=0,1,2,\ldots,(n+1)$, let $t \geq 0$ and let $\frac{|t|}{t} \min_i\{\text{Re}(\alpha_i)\} > a|t|$; then

$$\text{(i)} \quad |E^p_{MN} f| \leq \sup_{0 \leq \lambda < \infty} |e^{-a\lambda} f^{(n+1)}(\lambda)| \int_0^\infty |e^{a\lambda} K_{MN,n}(\lambda,t)| d\lambda$$

for $a=\sigma$

$$\text{(ii)} \quad |E^p_{MN} f| \leq \{\int_0^\infty |e^{-a\lambda} f^{(n+1)}(\lambda)|^p d\lambda\}^{1/p}$$
$$\{\int_0^\infty |e^{a\lambda} K_{MN,n}(\lambda,t) d\lambda|^q\}^{1/q}$$

$p>1$, $\frac{1}{p} + \frac{1}{q} = 1$, $a>\sigma$ .

Assertion 9

If $M=N-1$, then for every $N$ $\min_i\{\text{Re}(\alpha_i)\} > 0$.

The properties stated in Theorems 2,4,5 and 6 are similar to those of Padé approximants.

References

Zakian V., (1971). Electron. Lett., 7, pp.546-548.

Zakian V., (1972). Control Systems Centre Report No. 188, U.M.I.S.T., Manchester.

CRITICAL PHENOMENA AND PADE APPROXIMANTS

GENERALISED PADE APPROXIMANT BOUNDS FOR CRITICAL PHENOMENA*

George A. Baker, Jr.

(Applied Mathematics Department, Brookhaven National Laboratory, Upton, N.Y. 11973, U.S.A.)

In the early part of this lecture we will review the properties of the generalised Padé approximant (Baker, 1970) with an eye to using them to construct convergent, bounding successive approximations to several thermodynamic properties of some mathematical models of critical phenomena. Since Fisher has treated at some length the topic of critical phenomena at the preceding school and in the preceding lecture, I shall assume that the audience has by now gained enough background in that area.

As we have heard in previous lectures, the Padé approximant is extremely useful in providing bounds to functions of the class called series of Stieltjes defined by

$$f(z) = \int_0^\infty \frac{d\phi(t)}{1+zt} \qquad (1)$$

where $d\phi \geq 0$. The Padé approximation can be thought of as approximating $d\phi$ by a sum of $\delta$ functions. We wish to generalise Eq. (1) to

$$g(z) = \int_0^\infty b(z,s)\, d\phi(s) \qquad (2)$$

where $b(z,s)$ is now a more general function than the corresponding kernel in Eq. (1). We will consider again approximations such that $d\phi$ is approximated by a sum of $\delta$ functions. Thus we introduce the approximations to g(z)

$$B_{N,J}(x) = \sum_{m=1}^{N} \alpha_m b(x,\sigma_m) + \sum_{k=0}^{J} \frac{\beta_k}{k!} \left[\left(\frac{\partial}{\partial s}\right)^k b(x,s)\Big|_{s=0}\right]. \quad (3)$$

We obtain the [N+J/N] Padé approximant, $J \geqslant -1$, as a special case of Eq. (3) when we pick

$$b(x,s) = \frac{1}{1+xs} \quad (4)$$

as can be easily verified. As a matter of expository convenience let us introduce the notation

$$b(x,s) = \sum_{n=0}^{\infty} b_n(x)(-s)^n . \quad (5)$$

The approximations $B_{N,J}$ are determined by equating the low order coefficients in the expansion in powers of s. We assume that the $b_n(x)$ are such that these coefficients may always be determined.

Several theorems analogous to those for series of Stieltjes may be proved about this type of generalised Padé approximant. We give some of them following.

Theorem 1 If b(x,s) is regular in a uniform neighbourhood of the positive real s axis and $(\log s)^{1+\eta} b(x,s)$ is bounded as $s \to +\infty$ for some $\eta > 0$, then the approximants $B_{N,J}(x)$ converge as $N \to \infty$ for g(z) of the above form.

Proof Consider

$$\frac{1}{2\pi i} \int_C b(x,s) [N+J/N](-1/s)ds/s \quad (6)$$

where the contour C is the locus of all those points at a distance $\delta$ from the positive real axis in the s plane. Now by known theorems the [N+J/N] converges at every point of C and is bounded and analytic on C. We can put all this information together and prove convergence for all $s<S$ and, by the assumptions in the theorem, bound the integral over that part of the contour for

$s>S$ by $(\log S)^{-\eta}$. These facts suffice to prove the theorem.

Theorem 2 The $B_{N,J}(x)$ to above form satisfy

$$(-1)^{1+J}\{B_{N+1,J}(x) - B_{N,J}(x)\} \geq 0 \qquad (7a)$$

$$(-1)^{1+J}\{B_{N,J}(x) - B_{N-1,J+2}(x)\} \geq 0 \qquad (7b)$$

$$B_{N,0}(x) \geq g(x) \geq B_{N,-1}(x) \qquad (7c)$$

and $J \geq -1$ if and only if

$$\left(-\frac{\partial}{\partial s}\right)^{j} b(x,s) \geq 0 \qquad \begin{array}{l} 0 \leq x \leq \infty \\ 0 \leq s \leq \infty \\ j=0,1,2,\ldots \end{array} \quad . \qquad (8)$$

Proof The method of proof of this theorem is to show that under the conditions of the theorem suitable differences between successive $B_{N,J}(x)$ can be represented as function dependent constants of known sign times high order derivatives of $b(x,s)$ with respect to s. By an induction argument the conclusions of theorem can be established.

Theorem 3 (Bernstein) The statement

$$\left(-\frac{\partial}{\partial s}\right)^{j} f(s) \geq 0 \qquad (9)$$

and that $f(s)$ is regular for the real part of $s \geq 0$ and $j=0,1,2,\ldots$ is equivalent to

$$f(s) = \int_0^\infty e^{-st}\, d\phi(t) \qquad d\phi \geq 0 \ . \qquad (10)$$

Proof The proof of this theorem follows easily from the observation that a sufficiently high order derivative for a sufficiently large value of s can represent a delta function, almost, inside the integral form (10). Thus selecting a particular point t and via (9) we can show that the value of $d\phi$

must be positive at that point. That the integral form (10) implies (9) follows immediately by differentiation under the integral sign.

Let me list briefly some examples of suitable generalised kernels. Clearly $e^{-st}$ is an acceptable kernel, also the Leroy functions, $L_\zeta(st)$ where

$$L_\zeta(x) = \sum_{n=0}^{\infty} \frac{\Gamma(1+\zeta n)}{\Gamma(1+n)} \cdot (-x)^n = \int_0^\infty e^{-t-xt^\zeta} dt$$

are acceptable kernels. They have the property that they interpolate between for $\zeta=1$, the Padé type kernel and for $\zeta=0$, an exponential type kernel, by means of a family of entire functions (except for $\zeta=1$). Also clearly any derivative of an acceptable kernel is also an acceptable kernel. Hence differentiating the Padé kernel we obtain

$$(1+st)^{-n} \tag{11}$$

is also an acceptable kernel. It is worth noting that the so-called Borel-Padé summation procedure that is to say, combining both the Borel summation and the Padé approximant leads to the kernel

$$\int_0^\infty \frac{e^{-t}dt}{1+szt} \tag{12}$$

which is also an acceptable kernel for the properties of these theorems.

With this brief review of the properties of generalised Padé approximants we will turn to their application to the ferromagnetic Ising-Heisenberg model (Baker, 1971). For that system, the Hamiltonian is

$$H = - \sum_{i>j} J_{ij} [S_i^z S_j^z + \gamma_{ij} S_i^x S_j^x + \delta_{ij} S_i^y S_j^y] + mH \sum S_i^z \tag{13}$$

If we restrict

$$J_{ij} = J_{ji} \geqslant 0$$

$$\gamma_{ij} = \gamma_{ji} , \quad \delta_{ij} = \delta_{ji} , \quad |\gamma_{ij}| \leqslant 1 , \quad |\delta_{ij}| \leqslant 1$$

then it has been shown by Lee and Yang, (1952), Asano (1970), Suzuki (1968), (1969), Griffiths (1969) and Suzuki and Fisher (1971), that the partition function

$$Z = \sum_{\{\vec{S}_i\}} e^{-\beta H} , \quad Z = \prod_{i=1} (1-\mu/\mu_i) , \quad |\mu_i| = 1 \qquad (14)$$

has all its zeros, as a function of

$$\mu = \exp(-2m\ H/kT) , \qquad (15)$$

on the unit circle.

As Fisher has pointed out, the central problem in this field is to determine the critical behaviour at $T_C$, the temperature above which the spontaneous magnetisation is zero.

One can derive by standard methods that the spontaneous magnetisation is given by

$$I(\mu)/Nm = 1 - 2\mu \frac{d}{d\mu} \ln Z \qquad (16)$$

$$= \sum_{i=1}^{M} \frac{(1+\mu/\mu_i)}{(1-\mu/\mu_i)} . \qquad (17)$$

If $|\mu| \leqslant 1$ it follows readily from Eq. (17) that $\mathrm{Re}\{I(\mu)/Nm\} \geqslant 0$. Therefore by standard arguments it follows that

$$F(w) = (1+w)^{-\frac{1}{2}}\ I\left(\frac{(1+w)^{\frac{1}{2}}-1}{(1+w)^{\frac{1}{2}}+1}\right) \qquad (18)$$

is a series of Stieltjes for all temperatures. One can also easily show starting with the free energy

$$F/kT = -mH/kT - \int_0^1 \ln\left[(1-\mu)^2 + 4\mu y\right]\, d\phi(y) \qquad (19)$$

where $d\phi \geqslant 0$, because $d\phi$ represents the density of the zeros of the partition function, that it has an allowed kernel over the allowed range of the variables. Hence from the power series coefficients alone when available, we may give converging upper and lower bounds on the free energy.

Instead of the high field variables mentioned in the preceding analysis we seek results also in terms of low field variables for which series expansions are more often available. To this end, let us define

$$\tau = \tanh\left(\frac{mH}{kT}\right) . \qquad (20)$$

With some manipulation we may rewrite the formula for the spontaneous magnetisation as

$$G(\tau^2) = \frac{(I/mN)-\tau}{\tau(1-\tau^2)} = \int_0^\infty \frac{d\psi(\omega)}{1+\tau^2\omega} \qquad (21)$$

where $d\psi \geqslant 0$. When $T > T_C$, the upper limit of the integral in Eq. (21) is finite. However, for $T \lesssim T_C$, the upper limit of the integral is infinite. If we expand

$$G(\tau^2) = G_0(T) - G_1(T)\,\tau^2 + G_2(T)\,\tau^4 + \ldots \qquad (22)$$

and define the determinants

$$C(m+n/n+1) = \det \begin{vmatrix} G_m & \cdots & G_{m+n} \\ \vdots & \ddots & \vdots \\ G_{m+n} & \cdots & G_{m+2n} \end{vmatrix} \geqslant 0 \qquad (23)$$

which must be positive from the theorems then as

$$G_m \propto (T-T_C)^{-\gamma_m} \quad T \to T_C^+ \tag{24}$$

we must have the inequalities

$$\gamma_{i+1}-2\gamma_i+\gamma_{i-1} = \Delta_{i+1}-\Delta_i \geqslant 0 \tag{25}$$

in Fisher's notation. The scaling laws require the inequalities to be equalities. These theorems also have an experimental consequence, namely

$$(-1)^n \, \Delta^n_{\tau^2} \, G(\tau^2) \geqslant 0 \, . \tag{26}$$

It will be very interesting to see if this relation does in fact hold experimentally for real systems.

Next we show how the ideas of Villani (1972) and Fogli et al. (1971) can be combined with those of the generalised Padé approximant to give converging upper and lower bounds for the free energy per spin in the ferromagnetic Ising model (Baker, 1972). We will just sketch the outline here. The results will follow by summing, by generalised Padé procedures, a series whose coefficients depend on a parameter. In the physically interesting case all the series coefficients are infinite.

For convenience, we will introduce our Hamiltonian as

$$\mathcal{H} = \sum_{i,j} J_{ij}(1-\sigma_i\sigma_j) - mH \sum \sigma_i \tag{27}$$

where $J_{ij} \geqslant 0$ and the partition function is

$$Z = \sum_{\{\sigma_i=\pm 1\}} \exp\left[-\beta\mathcal{H}\right] \tag{28}$$

Because for any allowed state

$$\exp\left[-\beta J_{ij}(1-\sigma_i\sigma_j)\right] \leqslant 1 \tag{29}$$

always, it easily follows that

$$Z_{A \cup B} \leqslant Z_A \, Z_B \tag{30}$$

if A and B are disjoint sets of lattice sites. From (30) one can show that

$$\frac{1}{\#(2A)} \ln Z_{2A} \leqslant \frac{1}{\#(A)} \ln Z_A \tag{31}$$

where A is a set of the underlying lattice sites on which the spins are situated and 2A are two identical non-overlapping such sets. The function $\#(X)$ is the number of sites in set X. The conclusion of Eq. (31) is that the free energy (log partition function) per site is a monotonically decreasing function of the system size.

Next we quote our modified version of the Villani limit Theorem (Villani, 1972).

Theorem Let there be a sequence of functions $f_n(x)$ with the properties

(a) $f_n(1) = 2, \quad \lim_{x \to \infty} f_n(x) = 2, \quad f_n(x) < 2, \quad 1 < x < \infty$

(b) $\lim_{n \to \infty} f_n(x) = f(x), \quad 1 \leqslant x < \infty, \quad \lim_{x \to \infty} f(x) = \Lambda < 2$

(c) $f(x_1) \leqslant f(x_2), \quad x_1 \geqslant x_2$

(d) $f_n(x) \geqslant f(x), \quad 1 \leqslant x < \infty$

then there exists an infinite sequence of minima of $f_n(x)$, $x_n$ such that

$$\lim_{n \to \infty} x_n = \infty \tag{32}$$

$$\lim_{n \to \infty} f_n(x_n) = \Lambda \tag{33}$$

where by properties (c) and (d) $f_n(x_n) \geqslant \Lambda$ for all n.

The line of argument in the proof can best be understood by drawing the relationship between f(x) and $f_n(x)$. We have shown this relationship in Fig. 1. The idea is that as $f_n$ approximates f more closely, there will be a minimum for larger and larger x so that we get (32) since f is monotonic by (c). By the properties given in the theorem one can then demonstrate conclusion (33) also. The important thing to recognise about the assumptions of this theorem is the strong resemblance they bear to the results we have reviewed for generalised Padé approximants.

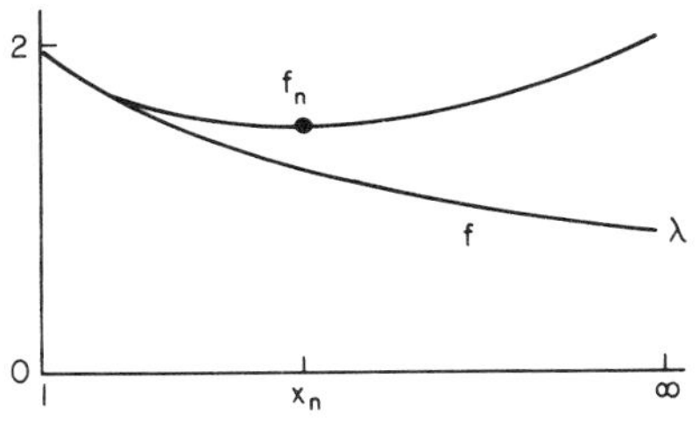


Fig. 1

The relationship between the various quantities in the Villani limit Theorem.

The final step is to exhibit the specific application of these results to the statistical mechanical problem at hand. We may rewrite Eq. (28) for the partition function as the Stieltjes integral

$$Z_N = \int_0^{\Gamma N} e^{-\beta E}\, d\rho_N(E) \qquad (34)$$

where N is the system size, ΓN is an upper bound on the total energy in any state, and $d\rho_N \geq 0$ as it is a density of states. Now

Eq. (34) is of the form (2) with an exponential kernel. Thus we expect to be able to apply generalised Padé procedure to (34) as a function of β. In particular, we find

$$\lim_{n\to\infty} {}_N B_{n,o}(\beta) = Z_N(\beta) \tag{35}$$

$${}_N B_{n,o}(\beta) \geqslant Z_N(\beta)$$

where the subscript N refers to the system size. If we now define

$$f_n(N) = [{}_N B_{n,o}(\beta)]^{1/N}, \quad f(N) = [Z_N(\beta)]^{1/N}$$

$$\Lambda(\beta) = \lim_{N\to\infty} [Z_N(\beta)]^{1/N} \tag{36}$$

then they satisfy all the conditions of the limit theorem since we have proved monotonicity of f. The allowed values of N are restricted to $2^{nd}$ (d = dimension) however it is reasonable to suppose that for uniform systems at least that this restriction is removable. Condition (a) follows because one can bound, for large N,

$$2^N \, c_n(\beta) \, N^{-n(n+1)} \leqslant {}_N B_{n,o}(\beta) \leqslant 2^N \tag{37}$$

so the $N^{th}$ root gives the right value as N goes to infinity.

We can also give convergent lower bounds by considering

$$\hat{H} = - \sum_{\substack{\text{nearest}\\ \text{neighbours}}} J(1+\sigma_i\sigma_j) - mH \sum_i \sigma_i \tag{38}$$

instead of (27), it is easy to show

$$\lim_{N\to\infty} [\hat{Z}_N(\beta)]^{1/N} = e^{-q\beta J} \, \Lambda(\beta) \tag{39}$$

where q is the lattice coordination number. But for $\hat{Z}$ the fundamental inequality (30) is reversed as now

$$\exp[\beta J(1+\sigma_i\sigma_j)] \geqslant 1 \tag{40}$$

A slightly more involved argument is now required as we do not have $\lim_{x\to\infty} f_n(x) = 2$, and energy $\hat{B}$ is evaluated for negative rather than positive values of its argument. In this case energy $\hat{B}$ is a lower bound. Consequently we conclude

$$\lim_{n\to\infty} \left[{}_{N(n)}\hat{B}_{n,j}(\beta)\right]^{1/N(n)} = \Lambda(\beta)\ e^{q\beta J}$$

$$\left[{}_{N(n)}\hat{B}_{n,j}(\beta)\right]^{1/N(n)} < \Lambda(\beta)\ e^{q\beta J} \tag{41}$$

where j=0,-1 are best for a fixed number of coefficients, and N(n) here is determined by maximisation.

Thus we have obtained rigorous, converging upper and lower bounds for all non-zero temperatures based on the high temperature series expansion coefficients in a field. By standard thermodynamics, we can generate convergent, but not necessarily bounding, sequences for any of the various thermodynamic variables.

As a simple illustration, we will use the linear chain in zero field. If J is the exchange integral and we use $K=\beta J$ for brevity, we have explicitly

$$Z_N = \sum_{\sigma_i=\pm1} \exp\left[-K \sum_{i=1}^{N-1} (1-\sigma_i\sigma_{i+1})\right] = 2(1+e^{-2K})^{N-1} \tag{42}$$

$$= 2^N\left[1-(N-1)K + \frac{1}{2} N(N-1)\ K^2 - \frac{1}{6}(N-1)^2\ (N+2)\ K^3 + \ldots\right.$$

and also

$$\hat{Z}_N = 2\ (1+e^{2K})^{N-1} \tag{43}$$

$$= 2^N\left[1+(N-1)K + \frac{1}{2} N(N-1)\ K^2 + \frac{1}{6}(N-1)^2\ (N+2)\ K^3 + \ldots\right. .$$

We compute the ${}_N B_{1,0}(\beta)$ and ${}_N\hat{B}_{2,-1}(\beta)$ bounds as

$$2e^{-2K}\left\{\frac{1}{2}\exp\left[(N-1-\sqrt{N-1})K\right]+\frac{1}{2}\exp\left[(N-1+\sqrt{N-1})K\right]\right\}^{1/N}$$

$$\leq (1+e^{-2K}) \leq 2\left[\frac{1}{N}+\left(1-\frac{1}{N}\right)e^{-NK}\right]^{1/N} \qquad (44)$$

For various values of K we get the results:

| K | Bounds | N,N |
|---|---|---|
| 0 | $2 \leq 2 \leq 2$ | 1,1 |
| 1 | $0.84 \leq 1.135 \leq 1.43$ | 16,4 |
| $\infty$ | $0 \leq 1 \leq 1.38$ | $\infty$,4 |

*Work supported in part by the U.S.A.E.C. and in part by A.R.P.A. through the Materials Science Centre of Cornell University.

References

Asano T., (1970). Phys. Rev. Let. 24, 1409.
Asano T., (1970). J. Phys. Soc. Japan 29, 330.
Baker G. A., Jr., (1970). In "The Padé Approximant in Theoretical Physics" (G. A. Baker and J. L. Gammel, eds.), pp.1-39, Academic Press, N.Y.
Baker G. A., Jr., (1971). In "Critical Phenomena in Alloys, Magnets and Superconductors" (R. E. Mills, E. Ascher and R. I. Jaffee, eds.), pp.221-229, McGraw Hill, N.Y.
Baker G. A., Jr., (1972). J. Math. Phys., to be published.
Fogli G. L., Pellicoro M. F. and Villani M., (1971). Nuovo Cimento (11) 6A, 79.
Griffiths R. B., (1969). J. Math. Phys. 10, 1559.
Lee T. D. and Yang C. N., (1952). Phys. Rev. 87, 410.
Suzuki M., (1968). J. Math. Phys. 9, 2064.
Suzuki M., (1969). Prog. Theoret. Phys. 41, 1438.
Suzuki M. and Fisher M. E., (1971). J. Math. Phys. 12, 235.
Villani M., (1972). In "Cargèse Lectures in Physics", Vol.5 (D. Bessis, ed.), Gordon and Breach, Paris, in press.

# CRITICAL PHENOMENA - SERIES EXPANSIONS AND THEIR ANALYSIS

Michael E. Fisher
(Baker Laboratory, Cornell University, Ithaca, New York 14850)

Summary*

At the critical point of a system (ferromagnet or antiferromagnet, fluid, superfluid, binary alloy, etc.) the thermodynamic properties become singular (i.e. nonanalytic) functions of the variables (temperature, pressure, magnetic field, etc.). In general terms the task of theory is to calculate the characteristic critical exponents, $\alpha, \beta, \gamma, \ldots \lambda, \ldots$ which describe the asymptotic behaviour of a property $F(z)$ in the vicinity of its critical point $z_c$, according to

$$F(z) \sim (z_c - z)^{-\lambda}, \quad z \rightarrow z_c- . \qquad (1)$$

Definitions of the standard critical exponents and expositions of the phenomenological theories of their interrelations, in particular the so-called scaling theories, will be found in the reviews: Fisher (1965, 1967) and Kadanoff et al. (1967). Significant progress has resulted from consideration of lattice models, especially the Ising model of a magnet (or, equivalently the lattice gas model). For such models between 8 to 30 coefficients $a_n$ of appropriate power series expansions

$$F(z) = \sum_{n=0}^{\infty} a_n z^n , \qquad (2)$$

can typically be obtained by a variety of special methods (see Domb 1960, Fisher 1965, 1967).

---

* Professor Michael E. Fisher felt that the topics he wished to discuss in his invited paper are adequately reviewed in the current literature, and so only a summary of his lectures appears in these proceedings.

In the simplest case the $a_n$ are real and positive, and the physical critical point $z_c$ determines the radius of convergence of the expansion (2). The ratios of coefficients, $\mu_n = a_n/a_{n-1}$, are typically observed to vary as

$$\mu_n \simeq \mu_\infty[1 + (g/n) + \ldots] \text{ as } n \to \infty . \quad (3)$$

It follows that numerical estimates of $\mu_\infty$ and g yield estimates for the desired critical parameters according to

$$z_c = 1/\mu_\infty \quad \text{and} \quad \lambda = 1 + g \quad . \quad (4)$$

Analysis of the ratios along these lines was pioneered by Domb and his co-workers (Domb and Sykes, 1961, Fisher and Sykes, 1959); the developments and refinements of the approach have been reviewed by Fisher (1965, 1967) and, more recently, by Gaunt and Guttman (1973) and Hunter and Baker (1973). The reliability of the techniques can be tested against exact solutions for two-dimensional Ising models (Onsager 1944).

In more general circumstances (typically at low temperatures) the series coefficients $a_n$ behave irregularly both as regards sign and magnitude, as the result of nonphysical, usually complex, singularities $z_i$ lying closer to the origin than the physical critical point $z_c$. Following the original suggestion of Baker (1961), one normally then analyses the logarithmic derivative expansion

$$D(z) = (d/dz) \ln F(z) = \sum_{n=0}^{\infty} d_n z^n \simeq \frac{-\lambda}{z-z_c} \; (z \to z_c) \quad (5)$$

by forming direct Padé approximants. The poles of the approximants, specifically of the paradiagonal sequences, lying in the vicinity of the expected physical critical point, yield estimates of the critical point $z_c$. At the same time the corresponding residues provide estimates of the exponent $\lambda$. Such methods using Padé approximants have been extended and widely applied by Baker and others (e.g. Essam and Fisher 1963).

and have played an important role in the theory of critical phenomena, both by providing strikingly good comparisons with experimental data and by giving reliable and precise evidence for, or against, more abstract theoretical approaches. Details of the techniques used and of various applications are given in the reviews: Fisher (1965, 1967), Gaunt and Guttmann (1973) and Hunter and Baker (1973).

While Padé approximants have proved invaluable for series with irregular coefficients, the ratio method is often to be preferred for the analysis of regular series, since the convergence is smoother and its asymptotic behaviour can more readily be related to the nature of competing singularities. Such information for Padé approximants could significantly enhance their usefulness. Effective approximants for nonanalytic functions of two variables would also be valuable in the study of critical phenomena.

## REFERENCES

Baker, G.A., Jr. (1961), Phys. Rev. 124, 768.

Domb, C. (1960), Advanc. Phys. 9, Nos. 34 and 35.

Domb, C. and Sykes, M.F. (1961), J. Math. Phys. 2, 63.

Essam, J.W. and Fisher, M.E. (1963), J. Chem. Phys. 38, 802.

Fisher, M.E. "The Nature of Critical Points" in "Lectures in Theoretical Physics," VIII C (Univ. of Colorado Press, Boulder) pp. 1 - 159.

Fisher, M.E. (1967), Reports Prog. Phys. 30, 615: especially sec. 7.

Fisher, M.E. and Sykes, M.F. (1959), Phys. Rev. 114, 45.

Gaunt, D.S. and Guttmann A.J. (1973) "Series Expansions: Analysis of Coefficients" in "Phase Transitions and Critical Phenomena" Vol. 3; Eds. C. Domb and M.S. Green, (Academic Press, London, New York).

Hunter, D.L. and Baker, G.A. Jr., (1973), "Methods of Series Analysis I : Comparison of Current Methods used in the Theory of Critical Phenomena" (in press).

Kadanoff, L.P. et. al. (1967), Rev. Mod. Phys. 39, 395.

Onsager, L. (1944), Phys. Rev. 65, 117.

A NEW METHOD OF SERIES ANALYSIS

G.S. Joyce[†] and A.J. Guttmann[*]

([†]Wheatstone Physics Laboratory, King's College, London, England)

([*]School of Mathematics, University of Newcastle, Australia)

In the lattice statistical theories of critical phenomena the usual procedure for investigating the singularities of a given thermodynamic function $\psi(T)$ is to calculate the coefficients $c_0, c_1 \ldots c_N$ in the Taylor series

$$\psi(z) = \sum_{n=0}^{\infty} c_n z^n , \qquad (|z| < r_0, \; c_0 \neq 0) \qquad (1)$$

where $z = z(T)$ is a suitable high or low temperature expansion variable. The singular behaviour of the series (1) and its analytic continuations in the z-plane are then determined approximately by applying various series analysis techniques, such as the ratio method (Domb and Sykes 1957) and the Padé approximant method (Baker 1961), to the truncated series $\sum_{n=0}^{N} c_n z^n$ Our main purpose in this paper is to outline a new general method of series analysis.

In this method the available series coefficients $c_0, c_1 \ldots c_N$ are fitted to a recurrence relation of the form

$$R_K(M_0 \ldots M_K; c_n) \equiv \sum_{i=0}^{K} \sum_{j=0}^{M_i} Q_{i,j} \, (n-j)^i \, c_{n-j} = 0, \; (n \geq 1) \qquad (2)$$

where $K \geq 0$, $M_i \geq 1$ $(i = 0,1 \ldots K)$, $Q_{K,0} \equiv 1$, $Q_{0,0} \equiv \delta_{K,0}$ and $c_{-n} \equiv 0$ . For fixed values of $K$ and $M_0 \ldots M_K$ , the set of coefficients $\{Q_{i,j}\}$ is calculated by solving the set of recurrence relations $\{R_K(M_0 \ldots M_K; c_n) = 0, \; n = 1 \ldots (\Theta + K - 1 + \delta_{K,0})\}$, where $\Theta = M_0 + \ldots + M_K$. This procedure is repeated, with $K$ fixed, for all values of $M_0 \ldots M_K$ which satisfy $(\Theta+K-1+\delta_{K,0}) \leq N$. In this manner we obtain a $(K+1)$-dimensional array of approximate recurrence relations for $c_n$.

Each recurrence relation (2) essentially <u>defines</u> a function $\psi_K[M_0 \ldots M_K;z]$ whose series expansion agrees with (1) through terms of order $\Theta+K-1+\delta_{K,0}$. Thus, we have

$$\psi(z) = \psi_K[M_0 \ldots M_K;z] + O(z^{\Theta+K+\delta_{K,0}}). \qquad (3)$$

We see, therefore, that the Kth order functions $\psi_K[M_0 \ldots M_K;z] \equiv \psi_K(z)$ provide us with an array of <u>approximants</u> for $\psi(z)$. Estimates for the singular behaviour of $\psi(z)$ are derived by investigating the appropriate singularities of $\psi_K(z)$. It is interesting to note that $\psi_0[M_0;z]$ is just the $[0/M_0]$ Padé approximant for the series (1).

For $K \geq 1$ it may be readily verified that $\psi_K(z)$ satisfies the Kth order linear differential equation

$$\sum_{i=0}^{K} Q_i(z)\, \Delta^i\, \psi_K(z) = 0, \qquad (\Delta \equiv z\,d/dz) \qquad (4)$$

where

$$Q_i(z) = \sum_{j=0}^{M_i} Q_{i,j}\, z^j, \qquad (i = 0,1\ldots K). \qquad (5)$$

This differential equation can be written in the alternative form

$$\sum_{\ell=0}^{K} z^\ell\, P_\ell(z)\, D^\ell\, \psi_K(z) = 0, \ (D \equiv d/dz) \qquad (6)$$

where

$$P_\ell(z) = \sum_{i=\ell}^{K} \bar{S}_i^{(\ell)}\, Q_i(z), \qquad (\ell = 0,1\ldots K) \qquad (7)$$

and $\bar{S}_i^{(\ell)}$ denotes a Stirling number of the second kind. We can now determine the nonanalytic behaviour of $\psi_K(z)$ by applying standard techniques to the differential equations (4) and (6).

When $K = 1$ the approximants $\psi_K(z)$ satisfy the equation

$$\left(\sum_{j=0}^{M_1} Q_{1,j} z^j\right) D\psi_1(z) + \left(\sum_{j=1}^{M_0} Q_{0,j} z^{j-1}\right) \psi_1(z) = 0 \qquad (8)$$

with $Q_{1,0} \equiv 1$ . This differential equation could also have been constructed by forming the $[M_0-1/M_1]$ Padé approximant to the logarithmic derivative of the series (1). We see, therefore, that the recurrence relation scheme for $K = 1$ is equivalent to the standard Padé approximant method of series analysis (Baker 1961).

We next consider the more interesting second order approximants $\psi_2[M_0,M_1,M_2;z]$ . In order to simplify the discussion we shall assume that the zeros $z_\alpha(\alpha = 1\ldots M_2)$ of the polynomial $P_2(z) \equiv Q_2(z)$ are all distinct, and that $P_\ell(z)$ , $(\ell = 0,1,2)$ have no common factors. (These conditions will in fact be satisfied in most numerical calculations.) Under these circumstances equation (6) has $M_2+1$ regular singular points in the finite z-plane at $z = z_0 \equiv 0$ , and $z = z_\alpha(\alpha = 1\ldots M_2)$. If $M_2 \geq M_i$ $(i = 0,1)$ , the point at infinity is also a regular singular point, and equation (6) becomes a Fuchsian differential equation. The exponents $\rho_\alpha,\xi_\alpha$ at each regular singular point $z_\alpha$ are calculated by solving the corresponding indicial equation. We find $\xi_\alpha = 0$ , with

$$\rho_\alpha = 1 - \lim_{z \to z_\alpha} (z - z_\alpha) \frac{P_1(z)}{zP_2(z)} , \quad (\alpha = 0,1\ldots M_2) . \tag{9}$$

(The singular point at $z = \infty$ requires a separate treatment.)

The behaviour of the approximant $\psi_2(z)$ in the neighbourhood of each regular singular point $z_\alpha$ is, in general, described by an analytic continuation of the form

$$\begin{aligned}
\psi_2(z) &= \phi_1(z) + (z_\alpha - z)^{\rho_\alpha}\phi_2(z), && (\rho_\alpha \neq 0, \pm 1, \pm 2, \ldots) \\
&= \phi_3(z) + \phi_4(z)\ln(z_\alpha - z), && (\rho_\alpha = 0) \\
&= \phi_5(z) + \varepsilon_1(z_\alpha - z)^{\rho_\alpha}\phi_6(z)\ln(z_\alpha - z), && (\rho_\alpha = 1,2,\ldots) \\
&= \{1 + \varepsilon_2\ln(z_\alpha - z)\}\phi_7(z) + (z_\alpha - z)^{\rho_\alpha}\phi_8(z), && (\rho_\alpha = -1,-2,\ldots)
\end{aligned} \tag{10}$$

where $\phi_i(z)$ are analytic and nonzero at $z = z_\alpha$ and $\varepsilon_1, \varepsilon_2$ are constants which are not necessarily different from zero.

In order to establish a link between equation (10) and the nonanalytic properties of $\psi(z)$ we assume that $\psi(z)$ has singularities at $z = \omega_i$ $(i = 1,2\ldots q)$, and that the structure of each singularity is given by (10) with an exponent $\lambda_i$ $(i = 1,2\ldots q)$. Under these circumstances it is reasonable to conjecture that a subset of $\{z_\alpha, \rho_\alpha; \alpha = 1\ldots M_2\}$ will yield approximations for the required set of positions and exponents $\{\omega_i, \lambda_i; i = 1\ldots q\}$, provided that $M_2 \geq q$. This conjecture forms the basis for the $K = 2$ recurrence relation method of series analysis. These arguments can be readily generalized for the higher order schemes $K = 3,4\ldots$ .

From the above discussion it is evident that the

recurrence relation method can deal directly with finite cusp singularities $(\lambda_i > 0)$, and with "weak" nonfactorizable divergent singularities $(\lambda_i \leq 0)$. Furthermore, the method can be used to analyse functions $\psi(z)$ which have several physical and nonphysical singularities on or outside the circle of convergence $|z| = r_o$. We see, therefore, that the recurrence relation technique is, at least in principle, more powerful than any other method of series analysis which is currently available. In addition, the method provides us with a mathematical tool for deducing exact recurrence relations and differential equations in lattice statistics and in the theory of special functions.

The recurrence relation method is currently being applied to a wide variety of series expansions in lattice statistics. We hope to discuss these applications in future publications.

Acknowledgment We are extremely grateful to Dr. M.F. Sykes for proposing various recurrence relation problems in the theory of the Ising model, from which the recurrence relation method evolved.

References

Baker, G.A. Jr., (1961) Phys.Rev. 124, 768.

Domb, C. and Sykes, M.F. (1957) Proc.Roy.Soc. A 240, 214.

# A COMPARISON OF THE VIBRATIONAL PROPERTIES OF H.C.P. AND F.C.C. CRYSTALS

C. Isenberg

(Physics Laboratory, University of Kent, Canterbury, England)

## 1. Introduction

Polymorphic transitions from a hexagonal close packed lattice structure, H.C.P., to a face centred cubic structure, F.C.C., are common. It is thus important to distinguish small differences in the thermodynamic properties of these similar structures.

In this paper the method of obtaining bounds for the thermodynamic properties, derived by Wheeler and Gordon (1969 and 1970), will be described, and used to compare the thermodynamic properties of the two lattices.

## 2. The Models

For simplicity the crystal models have only nearest neighbour central force interactions. An extension to models with short range forces can easily be made. The assumption of harmonic forces is also made.

The similarity of the two structures becomes clear if one considers the crystals to be constructed of triangular layers of atoms, as shown in the "projected" diagram, Fig. 1. The H.C.P. lattice is constructed by stacking triangular layers, vertically, at the positions indicated by layers A and B in the sequence ABABAB... The F.C.C. lattice has the arrangement ABCABCAB...

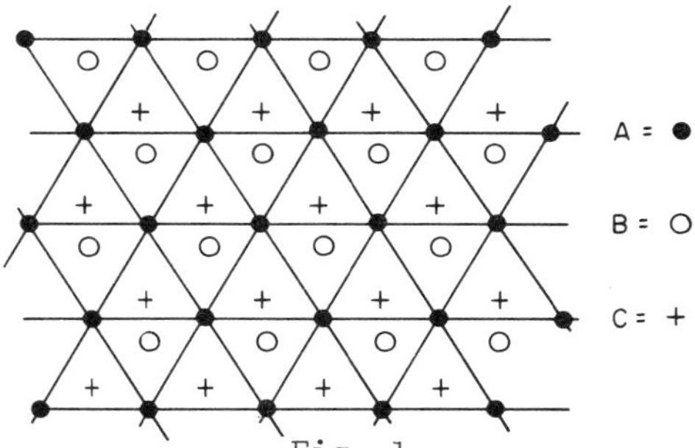


Fig. 1

## 3. Characteristics and Classification of Spectral Information

A typical vibrational spectrum has the qualitative features shown in Fig. 2. G(x) is the distribution of the squares of the normal mode frequencies, ω, and $x = \left(\frac{\omega}{\omega_{max}}\right)^2$, where $\omega_{max}$ is the maximum frequency.

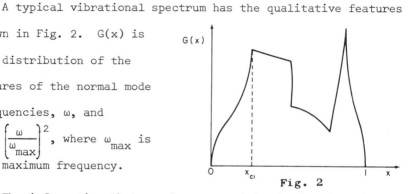

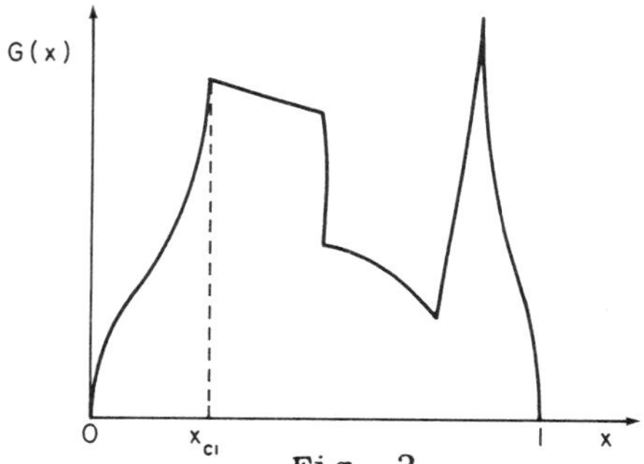


**Fig. 2**

The information that can be computed for both models is

a) Moments of the Frequency Spectrum

These are defined by

$$\mu_n = \int_0^1 x^n G(x)\, dx , \qquad n = 0, 1, 2, \ldots M$$

and can be computed exactly as rational numbers, Isenberg (1963 and 1970).

The first 19 moments have previously been obtained for the F.C.C. and 9 for the H.C.P. lattice, Isenberg (1971). Here the H.C.P. moments have been extended to 18 are are given in Table 1.

| n | $(48)^n \mu_n$ | n | $(48)^n \mu_n$ |
|---|---|---|---|
| 0 | 1 | 9 | 10048 31785 69560 |
| 1 | 24 | 10 | 4 25446 27788 98520 |
| 2 | 720 | 11 | 181 51256 02329 01152 |
| 3 | 24624 | 12 | 7794 70772 93281 45908 |
| 4 | 9 10728 | 13 | 3 36645 63712 60631 87544 |
| 5 | 353 35224 | 14 | 146 13676 88852 63663 64468 |
| 6 | 14136 90300 | 15 | 6373 09069 66010 52247 53024 |
| 7 | 5 77521 47232 | 16 | 2 79110 05701 07740 83900 57448 |
| 8 | 239 51496 11112 | 17 | 12271 42198 56992 46728 86732 968 |

Table 1 Moments of the H.C.P. lattice

b) Low Frequency Spectrum

The frequency spectrum G(x) can be expanded, for small x, to give the accoustical branches of the spectrum, Isenberg (1966), which determine the low temperature behaviour of the thermodynamic properties. This expansion has the form

$$2x^{\frac{1}{2}} G(x) = \sum_{r=1}^{\infty} A_r x^r , \qquad A_r > 0 \text{ for } r = 1, 2, \ldots$$

The expansion is convergent up to the first critical point $x_{CI}$. It is possible to compute the first R coefficients, R = O(20), with known accuracy. Consequently one can form a lower bound to the spectrum, $G^<(x)$, given by

$$2x^{\frac{1}{2}} G^<(x) = \sum_{r=1}^{R} A_r^< x^r , \qquad x \leqslant x_{CI}$$

where $A_r^< \leqslant A_r$ .

c) Critical Points and Critical Behaviour

The critical points arise at points of discontinuity in the gradient of the spectrum and can be determined exactly, Van Hove (1953).

## 4. Thermodynamic Properties

The thermodynamic properties $\mathfrak{F}(\tau)$, as a function of reduced temperature $\tau = \frac{kT}{\hbar\omega_{max}}$, can be determined from the relation

$$\mathfrak{F}(\tau) = \int_0^1 G(x)\, F(x,\tau)\, dx \,, \tag{4.1}$$

where $F(x,\tau)$ is the thermodynamic kernel. $\mathfrak{F}(\tau)$ can be expanded for large $\tau$ in a series whose coefficients are determined by the moments. The radius of convergence of the series is given by $\tau = \frac{1}{2\pi}$.

## 5. Moment Bounding Scheme

Gordon and Wheeler (1969 and 1970) have developed a method which enables accurate upper and lower bounds to be derived for the thermodynamic properties. Here the general method is described together with its modification to take account of the additional spectral information available. The method is equivalent to that given by Baker (1967) for bounding a series of Stieltjes.

If an even number, $M+1=2n$, of moments are available an upper bound can be constructed by the distribution

$$G^E(x) = \sum_{r=1}^{n} \omega_i \; \delta(x-x_i)$$

where the $\omega_i$'s and $x_i$'s are obtained by ensuring that the first (M+1) moments are obtained exactly.

$$\text{i.e.} \quad \sum_{i=1}^{n} \omega_i \, x_i^{\,k} = \mu_k \qquad\qquad k = 0, 1, \ldots M$$

$G^E(x)$ gives an upper bound to $\mathfrak{F}(\tau)$ when it replaces G(x) in (4.1).

Similarly, if an odd number, M+1=2n+1, of moments are available we can construct a distribution

$$G^O(x) = \sum_{i=0}^{n} \omega_i \; \delta(x-x_i) \qquad\qquad \text{where } x_o = 0$$

and whose first (M+1) moments are equal to $\mu_k$, k = 0, 1, ... M. This distribution provides a lower bound to $\mathfrak{F}(\tau)$.

The values of $\omega_i$ and $x_i$ can be computed in a well conditioned manner by using the product difference algorithm to form a tri-diagonal matrix whose eigenvalues and vectors give the $x_i$'s and $\omega_i$'s respectively.

This scheme can be extended to include additional spectral information. These improvements are

(i) Introduce (n+1) weighted δ-functions to form a distribution $G_R^E(x)$, for an even number of moments, with $x_o = 0$ and $x_{n+1} = 1$. Similarly form $G_R^O(x)$ with $x_{n+1} = 1$ and (2n+1) moments. Then either $G^E(x)$ and $G_R^E(x)$, or $G^O(x)$ and $G_R^O(x)$ form an improved set of upper and lower bounds to $\mathfrak{F}(\tau)$.

(ii) Use the lower bound to $G(x)$, $G^{<}(x)$, to form

$$\hat{G}(x) = G(x) - G^{<}(x) \qquad x \lesssim x_{CI}$$

$$= G(x) \qquad x > x_{CI} .$$

The moments for $\hat{G}(x)$, $\hat{\mu}_m$, are given by

$\hat{\mu}_m = \mu_m - \mu_m^{<}$, where $\mu_m^{<}$ are the exact moments formed from the low frequency spectrum $G^{<}(x)$. This improved scheme enables the contribution to $\mathfrak{F}(\tau)$ from $G^{<}(x)$ to be calculated exactly. The contribution from $G(x)$ can be bounded, as in (1), to give improved bounds to $\mathfrak{F}(\tau)$.

(iii) Finally $F(x,\tau)$ may diverge or have singular derivatives at $x = 0$ with the result that the bounds will go to infinity or be poor. Consequently it is often necessary to move the position of the $\delta$-function at $x = 0$, in the bounding distribution, to some non zero frequency so that bounds to $\mathfrak{F}(\tau)$ may be improved.

## 6. Conclusions

Although the work on bounding the thermodynamic properties of the H.C.P. and F.C.C. crystals is not complete, it is clear from the work of Wheeler and Gordon (1969) that the bounding scheme will enable the thermodynamic properties to be computed to an accuracy greater than previously possible. As an example, they have shown that the specific heat function, $C_v$, can be calculated over the whole temperature range to an accuracy greater than

$1:10^7$ for each model. The radius of convergence of the high temperature expansion is given by $\tau = \frac{1}{2\pi}$ .

References

Baker G. A. Jnr., (1967). Phys. Rev. 161, 434.

Isenberg C., (1963). Phys. Rev. 132, 2427.

Isenberg C., (1966). Phys. Rev. 150, 712.

Isenberg C., (1970). J. Phys. C. 3, L179.

Isenberg C., (1971). J. Phys. C. 4, 164.

Van Hove L., (1953). Phys. Rev. 89, 1189.

Wheeler J. C. and Gordon R. G., (1969). J. Chem. Phys. 51, 5566.

Wheeler J. C. and Gordon R. G., (1970). In "The Padé Approximant in Theoretical Physics" (G. A. Baker and J. L. Gammel, eds.), Academic Press, N.Y.

# ISING MODEL ANTIFERROMAGNETS WITH TRICRITICAL POINTS

F. Harbus and H. E. Stanley

(Massachusetts Institute of Technology, Cambridge, Massachusetts)

We have generated high-temperature series expansions to study the critical behaviour of two different Ising model Hamiltonians which exhibit tricritical behaviour. Analysing these series using the Padé approximant method extensively, we have determined the phase boundary in the field-temperature (H-T) plane and estimated values for the susceptibility exponents at the tricritical point.

Tricritical points, which occur in such physical systems as metamagnetic materials and liquid $He^3$-$He^4$ mixtures, are currently of considerable interest in the study of critical phenomena because entirely new critical behaviour, and, in particular, new critical point exponents, may be expected to characterise such points. This is due to the presence of two kinds of ordering processes occurring simultaneously. One is the second order phase transition associated with critical fluctuations in the order parameter. This transition is responsible for the line of singularities at high temperatures and low fields in metamagnetic antiferromagnets, where the order parameter is the staggered magnetisation $M_{st}$. At the tricritical point, however, fluctuations in the non-ordering density (the direct magnetisation M) become critical as well, signaling the onset of the first order transition occuring at low temperatures and higher fields.

To better appreciate the metamagnetic phase diagram, first consider the spin-1/2 Ising Hamiltonian ($s_i$=±1) for a simple 2 sublattice nearest-neighbour antiferromagnet on a simple cubic lattice,

$$H = -J_1 \sum_{ij} s_i s_j - \mu H \sum_i s_i . \qquad (1)$$

Here the exchange $J_1<0$ and acts isotropically in all lattice directions between nearest-neighbour (nn) spins. H is an external direct field. The field conjugate to the order parameter $M_{st}$, namely a staggered field $H_{st}$ acting in opposite senses on each sublattice, is not explicitly shown. The resulting phase diagram is expected to be as shown in Fig. 1.

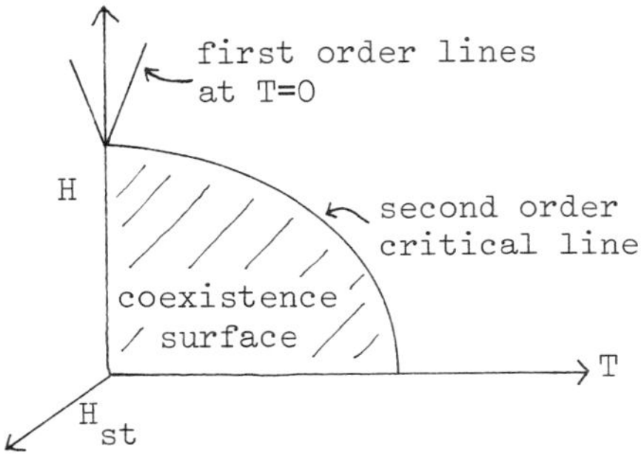


Fig. 1. Phase Diagram for Hamiltonian (1) in H-T-$H_{st}$ space.

There is a coexistence surface in the H-T ($H_{st}$=0) plane bounded by a <u>line</u> of critical points at which a second order transition associated with a divergent staggered susceptibility $\chi_{st}$ occurs. Only at T=0$^o$ is the transition first order, where first order "antennae" come out into the $H_{st}\neq 0$ region at 45$^o$ angles.

Now suppose the Hamiltonian (1) is complicated by adding <u>next</u> nearest-neighbour <u>ferromagnetic</u> interactions to give

$$H = -J_1 \sum_{ij} s_i s_j - J_2 \sum_{[ij]}^{nnn} s_i s_j - \mu H \sum_i s_i . \qquad (2)$$

The first term is as in (1), but the second has $J_2>0$ and is over next nearest-neighbour spins. It is the latter part of the Hamiltonian which gives rise to the appearance of a finite tricritical temperature $T_t$ and the first order phase transition behaviour below $T_t$. The phase diagram in the H-T plane now has a line of first order transitions whereas for (1) there existed in this plane only a lone first order point at T=0 as in Fig. 2.

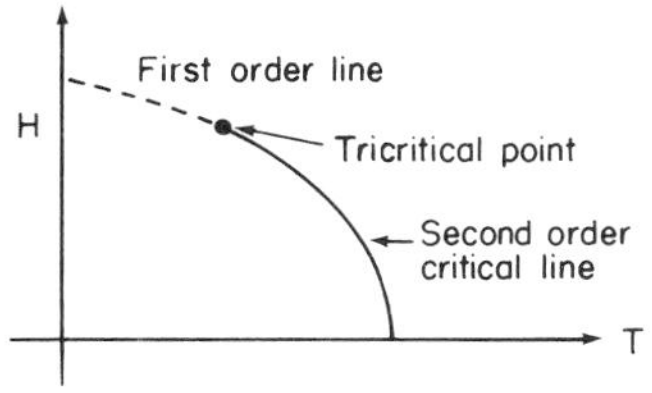


Fig. 2. Phase Diagram for Hamiltonian (2) in H-T Plane.

We may understand this behaviour physically by noting that the nnn term serves to stabilise the nn antiferromagnetic order at low temperatures, providing a "glue" which helps to lock in the order. In the presence of a critical external field entire blocks of spins locked together rather than spins one by one will flip as the system passes from the antiferromagnetic (AF) to paramagnetic phases.

The complete phase diagram in the 3-dimensional $H$-$T$-$H_{st}$ thermodynamic field space has been discussed recently by Griffiths (1970, 1971). It has coexistence surfaces or "wings" bounded by critical lines coming out into the $H_{st}\neq 0$ region in addition to the surface on the $H_{st}=0$ plane:

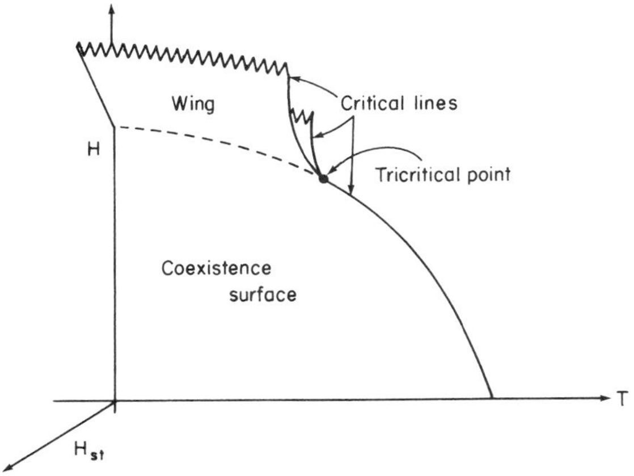


Fig. 3. Phase Diagram for Hamiltonian (2) in Complete $H$-$T$-$H_{st}$ space.

The first order line in the physical H-T plane lies at the intersection of the 2 wings, and the three critical lines meet at $T_t$.

A second Ising model with a tricritical point is described by a nn Hamiltonian but with different interactions in different lattice directions. Consider for the simple cubic lattice the Hamiltonian

$$H = -J_{xy} \sum_{ij}^{xy} s_i s_j - J_z \sum_{ij}^{z} s_i s_j - \mu H \sum_i s_i , \quad (3)$$

where the first sum is over nn spins in the x-y plane and the second over nn spins coupled in the z direction. To simulate metamagnetic materials such as $FeCl_2$ with in-plane ferromagnetic interactions and between-plane antiferromagnetic interactions, we take $J_{xy}>0$ and $J_z<0$. As in (2) the overall Hamiltonian is again antiferromagnetic in the sense that in the ordered state (consisting of a plane of spins all pointing up, the next plane all down, etc.) there is no net moment. The first term in (3) is the "glue" which locks spins within a given plane at low temperatures. We may now picture the external critical field as flipping whole portions of planes of spins aligned antiparallel to it, giving rise to the discontinuity in magnetisation observed experimentally below the tricritical temperature. It is perhaps more useful to think in terms of the order parameter $M_{st}$, which falls sharply to 0 along the 2nd order line but undergoes a discontinuous change below $T_t$.

The results for the model of Eq. (3) and some of the details of calculation are described elsewhere (Harbus and Stanley, 1972). We used identical methods to study the nnn model of Eq. (2) and here very briefly present results for it. Shown in Fig. 4 is the phase boundary resulting from high-temperature series expansions for the Hamiltonian (2) on the simple cubic lattice with the parameters $J_1=-1$, $J_2=+1/2$. The dots on the solid line are obtained from the $\chi_{st}$ series using Padé approximants and ratio techniques. The Padés continue to show singularities below the point estimated to be the tricritical point; the first order portion of the phase boundary is taken to join this point smoothly to the $T=0^{o}$ point, and is shown dashed. The consistency criterion that the Padé analysis of the series for $\chi$ produces poles in the tricritical region matching those from the $\chi_{st}$ series locates the tricritical point with greater precision, and further yields an estimate of 1/4 for the tricritical susceptibility exponent. Thus we conclude $\chi \sim (T-T_t)^{-1/4}$ for the nnn model.

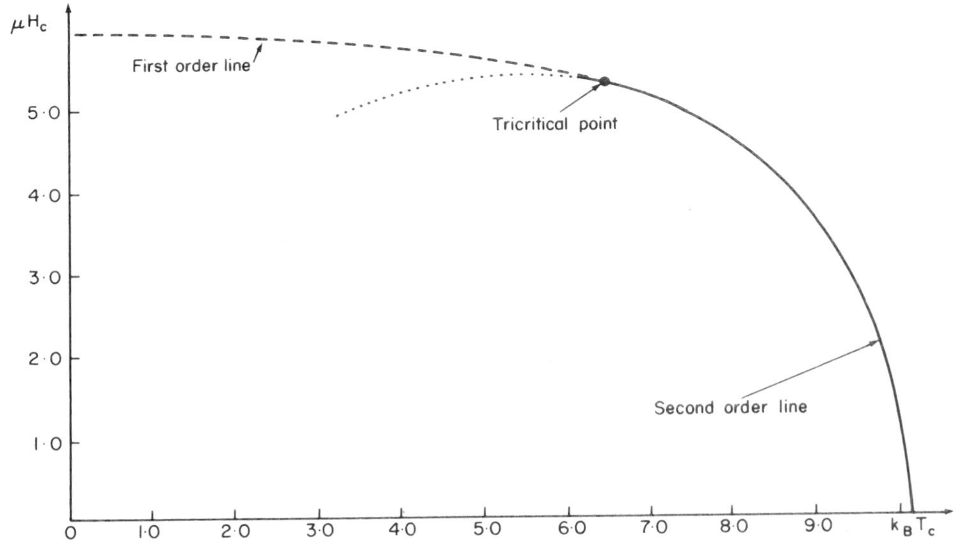


Fig. 4. Calculated Phase Boundary for simple cubic Ising Model (2) with $J_1=-1$, $J_2=+\frac{1}{2}$.

References

Griffiths R. B. (1970). Phys. Rev. Letters 24 715.

Griffiths R. B. (1971) in "Critical Phenomena in Alloys, Magnets and Superconductors" (R. E. Mills, E. Ascher and R. I. Jaffee, eds.) pp.377-391.

Harbus F. and Stanley H. E. (1972). Phys. Rev. Letters 29, 58.

# ISING, PLANAR AND HEISENBERG MODELS WITH DIRECTIONAL ANISOTROPY

F. Harbus, R. Krasnow, D. Lambeth, L. Liu and H. E. Stanley

(Massachusetts Institute of Technology, Cambridge, Mass., U.S.A.)

High-temperature series expansions have been an extremely useful tool in the investigation of the critical properties of various magnetic model Hamiltonians. A large number of these lattice models may be characterised by the Hamiltonian (Stanley, 1968a)

$$H = -J \sum_{\langle ij \rangle} \vec{S}_i^{(D)} \cdot \vec{S}_j^{(D)} \quad , \qquad (1)$$

where $\vec{S}_i^{(D)}$ is a D-dimensional classical vector situated on site i of the lattice, and the sum is over all nearest-neighbour pairs. Note D=1,2,3 correspond to the Ising, planar and classical Heisenberg models respectively, while D=∞ is the spherical model (Stanley, 1968b).

Whereas (1) has isotropic interactions with respect to lattice direction, a more general class of Hamiltonians (and undoubtedly truer approximations for real magnetic systems) would allow for different interaction strengths in different lattice directions. We focus upon the Hamiltonian

$$H = -J_{xy} \sum_{\langle ij \rangle}^{xy} \vec{S}_i^{(D)} \cdot \vec{S}_j^{(D)} - J_z \sum_{\langle ij \rangle}^{z} \vec{S}_i^{(D)} \cdot \vec{S}_j^{(D)}$$

$$\equiv -J_{xy} \left( \sum_{\langle ij \rangle}^{xy} \vec{S}_i^{(D)} \cdot \vec{S}_j^{(D)} + R \sum_{\langle ij \rangle}^{z} \vec{S}_i^{(D)} \cdot \vec{S}_j^{(D)} \right). \quad (2)$$

Now the first sum is over nearest-neighbour (nn) spins in an x-y plane, and the second sum is over nn spins coupled in the z direction.

Besides being better representations of real systems, the class of Hamiltonians (2) is of considerable theoretical interest because of its relevance to two concepts at the heart of much of our current understanding of critical phenomena. The first is the universality hypothesis which, in this context, predicts that critical behaviour for all systems with R>0 is ultimately 3-dimensional provided we are close enough to the critical temperature $T_c(R)$. That is, a model system with R>0 will be characterised by the appropriate 3-dimensional critical-point exponents, and only in the pure R=0 case do the exponents become 2-dimensional. Thus the universality class changes abruptly at R=0.

This change in universality class suggests that R becomes an "active" parameter for R close to zero, and ties in directly to the second important application of a high-temperature series study of (2). This is the testing of a generalised scaling hypothesis in the parameter R, an idea first developed by Riedel and Wegner (1969) in connection with a change in spin symmetry rather than lattice dimensionality.

To explicitly formulate the scaling hypothesis for the parameter R, we assume there exist 3 numbers $a_\tau$, $a_H$ and $a_R$ such that for all $\lambda>0$, the Gibbs potential obeys the functional equation

$$G(\lambda^{a_\tau}\tau, \lambda^{a_H}H, \lambda^{a_R}R) = \lambda\, G(\tau,H,R) \; . \qquad (3)$$

Here $\tau \equiv T-T_c(R=0)$ and H is the magnetic field.

From (3) may be derived predictions for the behaviour of the nth derivatives of thermodynamic functions derived from the Gibbs potential. For example, for the susceptibility $\chi$,

$$\chi^{(n)} \equiv (\partial^n \chi / \partial R^n)_{R=0} \sim [T-T_c(0)]^{-(\gamma_0+n\phi)} \quad , \qquad (4)$$

where $\gamma_0$ is the 2-dimensional susceptibility exponent and $\phi$ is the "crossover" exponent. In a similar fashion, one may make the (stronger) assumption that the pair correlation function $C_2$ scales in R, and have predictions regarding the behaviour of the second moment function $\mu_2 \equiv \sum_{\vec{r}} |\vec{r}|^2 C_2(\vec{r})$, where the sum is over all lattice sites.

We have obtained series coefficients (to tenth order in 1/T) as general polynomials in R. We treat the susceptibility, specific heat and second moment functions on the simple cubic and face centred cubic lattices for all three models (spin dimensionalities D=1,2 and 3).

The D=1 (Ising) case has been analysed in great detail using Padé approximants and other techniques. Our results confirm the prediction of scaling in R for both the thermodynamic functions and the pair correlation function; the details will be published elsewhere (Krasnow et al., 1972).

Similar analysis is in progress for the D=2 and 3 general-R series.

We present in Fig. 1 a sample of our D=3 data, the ratio plots for the first five derivatives of the second moment with respect to R on the fcc lattice for the Heisenberg model. The lines on this (and similar plots for other functions) appear to be pointing toward a non-zero critical temperature. The D=2 plots show analogous behaviour. These recent calculations seem to provide further evidence for a phase transition for a 2-dimensional lattice for both the planar and Heisenberg models.

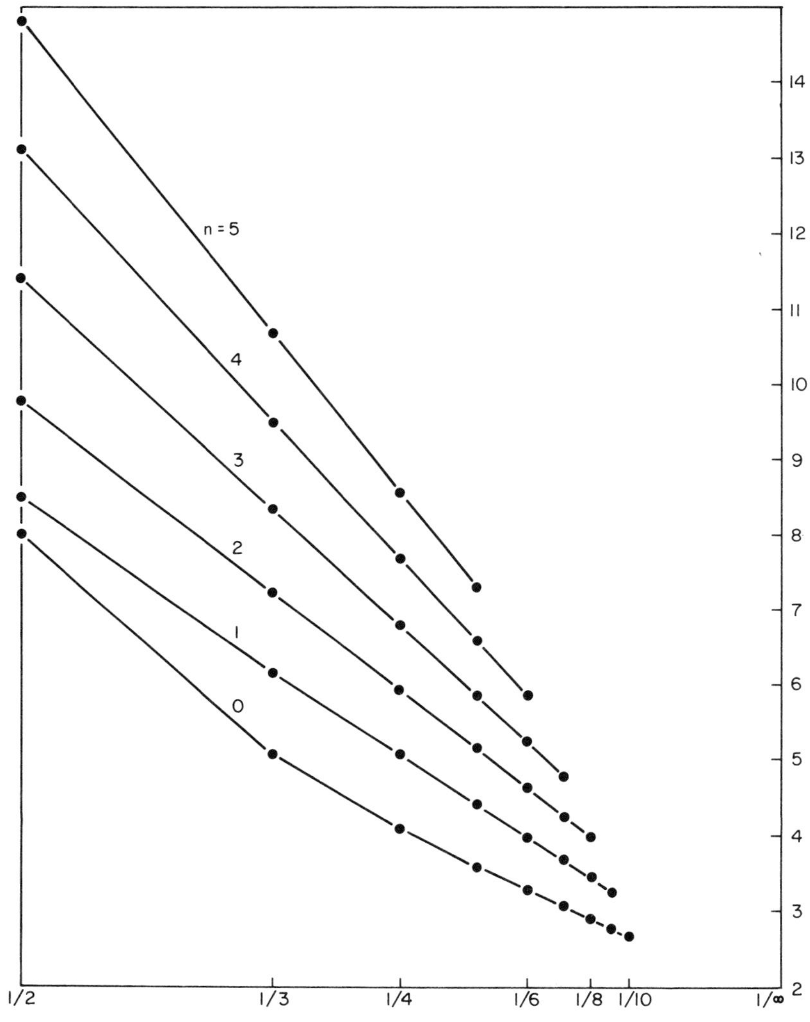


Fig. 1 The ratio plots for the first 5 derivatives of $\mu_2$ with respect to R for the fcc Heisenberg model.

References

Krasnow R., Harbus F., Liu L. and Stanley H. E., to be published.

Riedel E. and Wegner F., (1969). Z. Phys., 225, 195.

Stanley H. E., (1968a). Phys. Rev. Letters, 20, 589.

Stanley H. E., (1968b). Phys. Rev., 176, 718.

# CALCULATION OF THE EQUATION OF STATE NEAR THE CRITICAL POINT FOR THE HEISENBERG MODEL USING PADE APPROXIMANTS

Sava Milošević

(Institute of Physics, Belgrade, Yugoslavia)

and

H. Eugene Stanley

(Massachusetts Institute of Technology, Cambridge, Mass., U.S.A.)

The application of Padé approximants for constructing an equation of state near the critical point differs from the previous major utilisation of the same method for estimating critical point exponents. Whereas the critical point exponents can be, and indeed have been, estimated by using other approximation procedures, it turns out that the use of Padé approximants is inevitable in deducing the equation of state from the appropriate series expansion. Recently Gaunt and Domb derived the equation of state near the critical point for the Ising model, and their equation of state was expressed in the form of Padé approximants. We have calculated the equation of state for the $S=1/2$ and $S=\infty$ Heisenberg models, using Padé approximants, in an essentially different way from that which was done for the Ising model. We obtain a very satisfactory agreement with the available experimental data.

Before the appearance of the scaling hypothesis (Widom, 1965), (Domb and Hunter, 1965), and indeed for some time later, Padé approximants had been mainly applied as an estimation tool for the critical indices and related singular points (Stanley, 1971). When Gaunt and Domb (1970) undertook for the first time to calculate the whole equation of state for the Ising model, according to the scaling hypothesis, they found it possible to represent the equation of state by several Padé approximants. We have considered the application of Padé approximants in our calculation of the equation of state for the Heisenberg model as a means of representing results, in spite of the possible unphysical artifacts of the calculation procedure (Krasnow, 1972). We will discuss these artifacts here.

The static scaling hypothesis (Griffiths, 1967) suggests the following form of the equation of state near the critical point

$$H(\varepsilon,M) = M^{\delta}\, h(\varepsilon/M^{1/\beta}) \quad [M\geq 0] \qquad (1)$$

where H, $\varepsilon$ and M denote respectively singular parts of (i) the magnetic field divided by $kT_c/\mu$ ($\mu$ being the magnetic moment per spin), (ii) the reduced temperature $(T-T_c)/T_c$ and (iii) the magnetisation divided by its saturation value. The critical indices $\delta$ and $\beta$ describe the asymptotic behaviour of the critical isotherm and phase boundary respectively. The form of the function $h(x)$ is not specified by the scaling hypothesis, although it has certain well-defined characteristics (Griffiths, 1967). Invoking these characteristics, Gaunt and Domb (1970) calculated for the first time the $h(x)$ function for the Ising model. In the case of the Heisenberg model we could not follow Gaunt and Domb as some prerequisite information (the so-called low-temperature expansions) are not known for this model.

However, one can argue that the scaling hypothesis in the form (1) implies that $H(\varepsilon,M)$ is a generalised homogeneous function, i.e. it satisfies the relation

$$H(\lambda^{1/\beta(\delta+1)}\varepsilon,\ \lambda^{1/(\delta+1)}M) = \lambda^{\delta/(\delta+1)}H(\varepsilon,M)\ , \qquad (2)$$

where $\lambda$ is any positive parameter. Hence, choosing $\lambda$ to be

$(c/M)^{\delta+1}$, where the number c is assumed to be very small, (2) results in

$$c^{\delta} H(\varepsilon,M)/M^{\delta} = H(\varepsilon\, c^{1/\beta}/M^{1/\beta},\, c) \qquad (3)$$

and thereby one may further argue that

$$h(x) = H(xc^{1/\beta},c)/c^{\delta} \;. \qquad (4)$$

Thus it appears that h(x) is not a completely unknown function - if we know $H(\varepsilon,M)$ then h(x) can be obtained by replacing the variables $\varepsilon$ and M by $xc^{1/\beta}$ and c respectively. The function H has its values in the critical region providing its arguments are very small quantities and hence (4) is correct with the proviso that c is a very small quantity.

The most we know about the function $H(\varepsilon,M)$ for the Ising and Heisenberg models is a finite number of terms in its series expansion. Therefore in order to apply (4) for obtaining the function h(x) we must extrapolate the regular behaviour of these terms using the technique of Padé approximants.

In the case of the Ising model on the bcc lattice, the series for $H(\varepsilon,M)$ has the following form (Gaunt and Baker, 1970)

$$H(\varepsilon,M) = (\varepsilon+1)\ \mathrm{arcth}\left[M\tau(M,v)\right] \qquad (5)$$

where

$$\tau(M,v) = \sum_{n=o}^{L} \psi_n(M)\, v^n \;. \qquad (6)$$

Here $\psi_n(M)$ are polynomials in $M^2$ of degree n,

$$v \equiv \tanh(J/kT) \equiv \tanh\left[K_c/(\varepsilon+1)\right], \qquad (7)$$

$K_c=J/kT_c$ and J is the exchange parameter in the Ising Hamiltonian. In this case L=12 instead of being infinity (we simply do not know more polynomials $\psi_n$ for the time being).

According to (4), in order to obtain h(x) we must set M=c in (5) where c is to be a small positive constant. Formally this step is similar to the problem encountered by Gaunt and Baker (1970) in connection with their calculation of the phase boundary and we are going to follow their approach here. We assume that the function $\tau(M{=}c,v)$ vanishes at the phase boundary with the power law form

$$\tau(c,v) = \sum_{n=o}^{L} \psi_n(c)\, v^n = (1 - \frac{v}{v_o})^q\, f(v) \quad (8)$$

where $v_o$, q and f(v) are to be estimated by the method of Padé approximants. Thus one finds $(v_o,q)$ and f(v) by considering Padé approximants to $(d/dv)\,[\ln \tau(c,v)]$ and $(1-v/v_o)^{-q}\,\tau(c,v)$ respectively.

It was noticed (Gaunt and Baker, 1970) that the series (6) was not sufficiently lengthy for reliable estimates for $v_o$ and q to be obtained unless $c \gtrsim 0.6$. Since the smaller the values of c, the larger region of x where the relation (4) is satisfied, we chose c=0.6. Table 1 contains poles and residues of Padé approximants to $(d/dv)\,[\ln \tau(0.6,v)]$; these correspond to $v_o$ and q of (8) respectively. We estimated from Table 1 that $v_o$=0.1658 and q=1.076. Then we formed Padé approximants to the function f(v); these approximants were found to be consistent up to five decimal places (cf. Table 2). We therefore almost arbitrarily chose the [4,4] approximant, whence (8) becomes

$$\tau(0.6,v) = (1-v/v_o)^q \; \frac{1-4.566v+5.406v^2+5.842v^3+0.3907v^4}{1-5.936v+17.603v^2-37.098v^3+25.812v^4} \quad (9)$$

where $v_o$=0.1658 and q=1.076.

If we now combine (9), (8) and (5) we can obtain from (4) an expression for h(x)

$$h(x) = [(0.195x+1)/0.07776]\ \mathrm{arcth}(0.6\ \tau(0.6,\tilde{v})), \quad (10)$$

TABLE 1

| N \ M | 2 | 3 | 4 | 5 | 6 | 7 | 8 | 9 |
|---|---|---|---|---|---|---|---|---|
| 2 | 0.1847 (1.6788) | 0.1601 (0.8770) | 0.1638 (1.0052) | 0.1692 (1.3012) | 0.1665 (1.1215) | C.Z. | 0.1658 (1.0767) | 0.1658 (1.0804) |
| 3 | 0.1669 (1.1288) | 0.1662 (1.1016) | 0.1662 (1.1015) | 0.1661 (1.0973) | 0.1661 (1.0954) | 0.1659 (1.0868) | 0.1658 (1.0816) | |
| 4 | 0.1649 (1.0330) | 0.1662 (1.1014) | 0.1662 (1.1016) | 0.1660 (1.0939) | C.Z. | 0.1658 (1.0758) | | |
| 5 | 0.1662 (1.1033) | 0.1661 (1.0973) | 0.1660 (1.0911) | C.Z. | C.Z. | | | |
| 6 | C.Z. | 0.1660 (1.0930) | C.Z. | C.Z. | | | | |
| 7 | 0.1661 (1.0949) | 0.1658 (1.0688) | C.Z. | | | | | |
| 8 | C.Z. | C.Z. | | | | | | |
| 9 | 0.1658 (1.0794) | | | | | | | |

For each Padé approximant $[N,M]$ the upper number corresponds to $v_o$, while the number in brackets approximates $q_o$ of (8). Here C.Z. (competing zero) means that the Padé approximant erroneously predicts that $\tau(0.6,v)$ has two zeros, one below $T_o$ and the other above $T_c$.

where $\tau(0.6,\tilde{v})$ is given by (9) with $\tilde{v}=\tanh(0.15743/(0.195x+1))$, and we have used $\beta\simeq 5/16$ and $\delta\simeq 5$ (Stanley, 1971).

We have to emphasise that the expression (10) cannot represent $h(x)$ for large $x$. It is because $c=0.6$ is not sufficiently small to maintain the validity of relation (4) for relatively big $x$. For $x\gtrsim 1$, (10) has to be supplemented by one expression which follows from an expansion of $h(x)$ near the critical isochore (Griffiths, 1967). We found this supplementary expression in essentially the same way utilised by Gaunt and Domb (1970).

In the case of the S=1/2 Heisenberg model Baker et al. (1970) calculated high-temperature series expansions that are analogous to the Ising model series (5)

$$H(\varepsilon,M) = (\varepsilon+1)\ \text{arcth}[M\ g(M,z)] \qquad (11)$$

where

$$g(M,z) = \sum_{n=o}^{L} (2^{-n}/n!)\ P_n(M)\ z^n \qquad (12)$$

and $P_n(M)$ are polynomials in $M^2$ of degree $n$, $z=K_c/(\varepsilon+1)$ whereas $L=8$ for the fcc lattice which we will consider here. In analogy with (8) we assume that the function $g(M,z)$ for fixed $M=c$ vanishes at the phase boundary as

$$g(c,z) = (1 - \frac{z}{z_o})^q\ \phi(z)\ . \qquad (13)$$

The Padé approximant analysis of $(d/dz)\ [\ln g(c,z)]$ provides relatively reliable results for $z_o$ and $q$ providing $c$ is in the range $0.4\lesssim c\lesssim 0.85$. For the same reasons discussed in the case of the Ising model we choose the smallest possible value, $c=0.4$. Our estimates for $z$ and $q$ are essentially the same as those of Baker et al. (1970), $z_o=0.25526$ and $q=1.29$. The Padé approximants to the function $\phi(z)$ of (13) were found to be quite consistent and we chose the [3,3] approximant as representative. The corresponding closed-form expression for $g(0.4,z)$ is

$$g(0.4,z) = (1-z/0.25526)^{1.29} \frac{1+3.789z+1.671z^2+3.612z^3}{1+3.775z+4.622z^2+14.397z^3} . \tag{14}$$

Substituting (14) into (11) and recalling (4) we obtain the following expression for the scaling function $h(x)$ of the $S=1/2$ Heisenberg model

$$h(x) = \frac{(0.4)^{1/\beta}x+1}{(0.4)^{\delta}} \operatorname{arcth}(0.4\, g(\tilde{z},0.4)) , \tag{15}$$

where

$$\tilde{z} = 0.2492/((0.4)^{1/\beta}\, x+1) . \tag{16}$$

Again we must emphasise that (15) is valid for small x. In order to obtain an expression for large x we used the Gaunt-Domb method. We have also to stress that indices $\beta$ and $\delta$ which appear in (15) are not yet firmly determined, at least as definitely as in the Ising model case. There is an alternative (Milošević and Stanley, 1972) to use either $\beta=0.35$ and $\delta=5$ or $\beta=0.385$ and $\delta=4.71$. In Fig. 1 we present a comparison of the scaling functions for the Heisenberg model, calculated according to the both possibilities, together with the scaling function for the Ising model and experimental results for $CrBr_3$ (Ho and Litster, 1969). We have also calculated the $h(x)$ function for the $S=\infty$ Heisenberg model, but from the Padé approximant point of view it is quite sufficient to discuss the $S=1/2$ case.

One can notice from (10) and (15) that the $h(x)$ functions vanish when $\tilde{v}$ approaches 0.1658 and $\tilde{z}$ approaches 0.25526 respectively. In general $h(x)$ vanishes when x becomes $-x_o$, or in accord with (1) when $M=x_o^{-\beta}(-\varepsilon)^{\beta}$, i.e. it happens at the phase boundary when $H=0$. However, it can be conceived (Krasnow, 1972) from (10) and (15) that $h(x)$ has the following asymptotic form close to $-x_o$

$$h(x) = (x+x_o)^q\, \phi(x) , \tag{17}$$

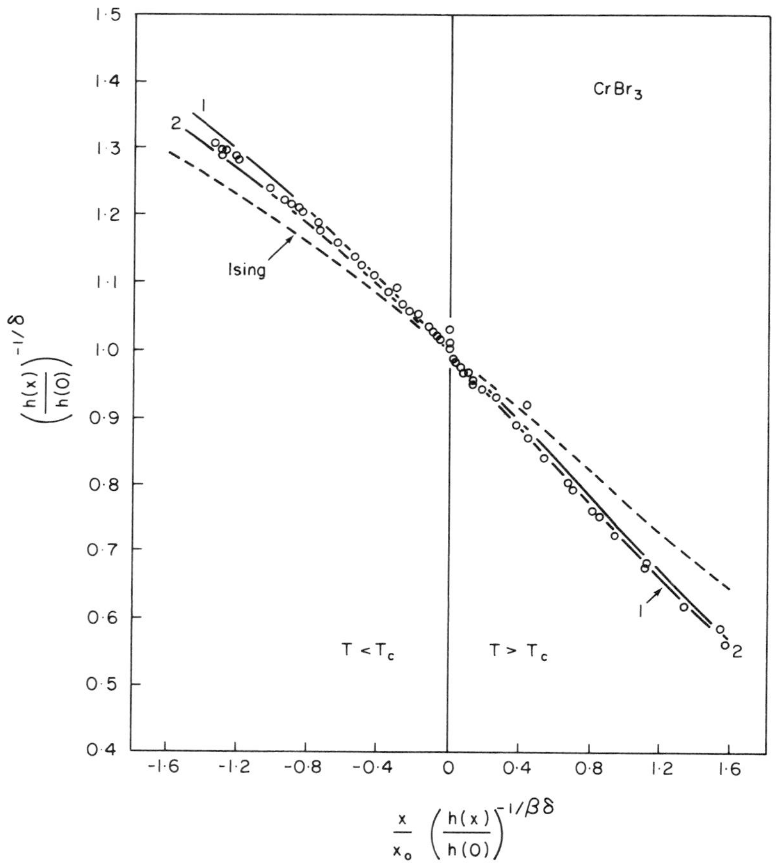


Fig. 1

Comparison of the scaling functions h(x) for the Ising and Heisenberg models with the experimental results for $CrBr_3$. The curve labelled 1 corresponds to the Heisenberg model with $\beta$=0.385 and $\delta$=4.71, whereas the curve numbered 2 corresponds to the same model with the values $\beta$=0.35 and $\delta$=5.

where the function $\phi(x)$ is free of singularity at $-x_0$. Hence the first derivative of $h(x)$ will vanish at $x=-x_0$ unless $q \lesssim 1$. Since $h(x)=H/M^{\delta}$ it means that the susceptibility $\chi=(\partial M/\partial H)_{H=0}$ will go to infinity at the phase boundary if $q>1$. As we found $q>1$ one can argue that this is a rather unphysical result (Krasnow, 1972). It remains to be pondered whether this result is an artifact of the calculational method, and particularly to be cleared up whether the assumptions (8) and (13) were reasonable.

In the process of determination of the phase boundary from the high-temperature series for the Ising model Gaunt and Baker (1970) introduced the exponent $\iota \equiv 1/q$ arguing that $\iota=1$ is almost certainly correct. Of course, we explored the possibility of using $q=1$. Table 2 provides values of $h(0)$ obtained from expressions fully analogous to (9) only constructed with different Padé approximants to the function $f(v)$ of (8). Column A is constructed using $q=1.076$, column B uses $q=1.08$ while column C uses the prediction $q=1$. One can notice in column A the rather striking consistency between different Padé approximants (as is typical for other values of $h(x)$ as well). It is noticeable from column B that this consistency is weakened when one takes a sort of average estimate $q=1.08$ from Table 1 for q. From column C we see that the consistency gets still worse when one takes $q=1$. The consistency of the Padé approximants results led us to use $q=1.076$ in (9) instead of the presumably correct value $q=1$. However, this may be merely a way to achieve a consistent approximation but not an attempt to disprove the possibility of q being equal to unity.

Baker et al. (1970) wondered whether $g(M,z)$ (of (11)) vanishes in a simple manner at the phase boundary or has, perhaps, a branch point there, i.e. whether $q=1$ or $q\neq 1$ in (13). In order to determine the phase boundary for the $S=1/2$ Heisenberg model the same authors found it convenient to assume that $q\neq 1$ and

TABLE 2

| Padé [N,M] | A (q=1.076) | B (q=1.08) | C (q=1) | D (q=1.076, δ=4.8) |
|---|---|---|---|---|
| [3,3] | 0.3812 | 0.3807 | 0.3911 | 0.3442 |
| [3,4] | 0.3819 | 0.3812 | 0.3906 | 0.3448 |
| [3,5] | 0.3819 | 0.3812 | 0.3889 | 0.3448 |
| [3,6] | 0.3819 | 0.3813 | 0.3874 | 0.3448 |
| [4,3] | 0.3819 | 0.3812 | 0.3906 | 0.3448 |
| [4,4] | 0.3819 | 0.3813 | 0.3914 | 0.3448 |
| [4,5] | 0.3819 | 0.3813 | 0.3769 | 0.3448 |
| [4,6] | 0.3819 | 0.3811 | 0.3856 | 0.3448 |
| [5,3] | 0.3819 | 0.3813 | 0.3899 | 0.3448 |
| [5,4] | 0.3819 | 0.3814 | 0.3812 | 0.3448 |
| [5,5] | 0.3819 | 0.3812 | 0.3868 | 0.3549 |
| [5,6] | 0.3930 | 0.3820 | 0.3842 | 0.3448 |
| [6,3] | 0.3819 | 0.3813 | 0.3888 | 0.3448 |
| [6,4] | 0.3819 | 0.3812 | 0.3863 | 0.3448 |
| [6,5] | 0.3819 | 0.3801 | 0.3846 | 0.3448 |
| [6,6] | 0.3819 | 0.3817 | 0.3830 | 0.3448 |

Values of h(o) for the Ising model obtained by using different Padé approximants (different rows of the table) to the function f(v) of (8), and by choosing different estimates of q.

thereby obtained a better convergence between the different Padé approximants. But they also did not find it possible unambiguously to determine $\iota \equiv 1/q$ from the series expansions. We were faced with the exactly same situation and decided to use that value of q which emerged from the relatively most consistent Padé table. It turned out to be q=1.29 when c=0.4 in (13). Precisely the same value was suggested by Baker et al. (1970).

Therefore for the Ising model as well as for the Heisenberg model we used q>1. The unphysical implications suggested by (17) are almost certainly inappropriate for the Ising model. In the case of the Heisenberg model it is not yet evident whether the unphysical predictions come inherently from the model, or, more likely, they are consequences of the limited number of terms in the series we dealt with. Hence we are tempted to consider (9) and (14) merely as numerical approximations to true equations of state, rather than suggestions of the possible forms these equations should have in the critical region.

## References

Baker G. A., Jr., Eve J. and Rushbrooke G. S., (1970). Phys. Rev. B2, 706.

Domb C. and Hunter D. L., (1965). Proc. Phys. Soc. (London) 86, 1147.

Gaunt D. S. and Baker G. A., Jr., (1970). Phys. Rev. B1, 1184.

Gaunt D. S. and Domb C., (1970). J. Phys. C3, 1442.

Griffiths R. B., (1967). Phys. Rev. 158, 176.

Ho J. T. and Litster J. D., (1969). Phys. Rev. Letters 22, 603.

Krasnow R., (1972). To be published.

Milošević S. and Stanley H. E., (1972). Phys. Rev. B6, 986 and 1002.

Stanley H. E., (1971). In "Introduction to Phase Transitions and Critical Phenomena", Oxford University Press, London.

Widom B., (1965). J. Chem. Phys. 43, 3898.

# ATOMIC, NUCLEAR AND PARTICLE PHYSICS AND PADE APPROXIMANTS

APPLICATIONS OF THE MOMENT PROBLEM

A.K. COMMON

(Mathematical Institute, University of Kent, England)

## 1. Introduction

This paper is a review of work dealing with a fruitful application of the moment problem in deriving constraints on scattering amplitudes from basic assumptions of crossing, unitarity and analyticity. The reason for discussing this work is not only to describe how the above constraints are derived but also to bring out aspects of the moment problem which may have useful applications elsewhere.

The above properties of the scattering amplitude, which hold in the framework of axiomatic field theory, are now described. For simplicity the process

$$\pi^o(p_1)\ \pi^o(p_2) \rightarrow \pi^o(-p_3)\ \pi^o(-p_4) \tag{1}$$

is considered, although the results can be extended to include processes involving charged pions and other particles.

Let the four pions have the momenta indicated and choose units so that $\hbar = c =$ 2 x mass of the pion = 1. Then the scattering amplitude is a function of two independent kinematic variables which are taken to be

$$s = (p_1 + p_2)^2 = (\text{C.M. energy})^2 \tag{2}$$

and

$$t = (p_1 + p_3)^2 = \tfrac{1}{2}(1-s)(1-\cos\Theta_s) \tag{3}$$

where $\Theta_s$ is the scattering angle in the cente of mass frame.

It is also usual to introduce the quantity

$$u = (p_1 + p_4)^2 = \tfrac{1}{2}(1-s)(1+\cos\theta_s), \qquad (4)$$

so that

$$s + t + u = 1$$

Even though u is determined by s and t it is useful to consider the scattering amplitude as a function of all three variables.

In the framework of axiomatic field theory, it may be proved that the scattering amplitude is the boundary value of a function $F(s,t,u)$, analytic in the whole complex s-plane cut from 1 to $\infty$ and $-t$ to $-\infty$ and for all $|t| < 1$ and in a similar domain with s and t interchanged, (Martin, 1963). The values $s \geq 1$ and $o \geq t \geq (1-s)$ correspond to physical process (1) and the scattering amplitude is taken to be the values of $F(s,t,u)$ as this boundary is approached from the upper half s-plane.

The values $t \geq 1$ and $o \geq s \geq (1-t)$ correspond to a process with particles 1 and 3 ingoing and 2 and 4 outgoing. Because this again is the scattering process $\pi^o\pi^o \to \pi^o\pi^o$, it may be proved that for all s and t in the analyticity domain

$$F(s,t,u) = F(t,s,u) \qquad (6)$$

and this result may be generalized to obtain invariance under all permutations of s,t and u.

Jin and Martin (1964) have shown that for $|t| \leq 1$,

$$|F(s,t,u)| \leq Cs^2. \qquad (7)$$

Therefore $F(s,t,u)$ satisfies a dispersion relation for $|t| \leq 1$, which using (6) and its generalization may be written in form,

$$F(s,t,u) = \alpha(t) + \frac{1}{\pi}\int_1^\infty \mathrm{Im}F(s'+i\varepsilon,t)\left|\frac{s^2}{(s'-s)s'^2} + \frac{u^2}{(s'-u)s'^2}\right| ds' \qquad (8)$$

Another consequence of the analyticity properties of F(s,t,u) is that for $s \geqslant 1$ and $|t| \leqslant 1$ it has the convergent partial wave expansion,

$$F(s,t,u) = \sum_{\ell=0}^{\infty} (2\ell+1)\, f_\ell(s)\, P_\ell(\cos \Theta_s). \tag{9}$$

The unitarity condition is that for physical values of s the partial - waves may be written in the form

$$f_\ell(s) = \frac{\sqrt{s}}{2iq} \,|\eta_\ell e^{2i\delta_\ell} - 1| \tag{10}$$

where $1 \geqslant \eta_\ell \geqslant o$ and $\delta_\ell$ are real and where q is the c.m. momentum.

From (10) it is immediately seen that

$$1 \geqslant \text{Im } f_\ell(s) \geqslant 0, \tag{11}$$

and that for $s \geqslant 1$ and $1 \geqslant t \geqslant 0$,

$$\text{Im } F(s + i\varepsilon, t) \geqslant 0 \tag{12}$$

since $P_\ell(\cos\Theta_s) \geqslant 1$. The conditions (11) and (12) are the consequences of unitarity which will be used in this work.

The above then are the basic assumptions on F(s,t,u) and they lead to constraints on the behaviour of the scattering amplitude in the physical region and also the unphysical region $0 \leqslant s,t,u \leqslant 1$. It is obvious that the latter set of constraints cannot be compared directly with experiment, but they do prove useful in the construction of models for the partial waves $f_\ell(s)$. It is fairly easy to construct a model for such a wave which satisfies the unitarity condition (10) but not the crossing condition (6) since the latter involves an infinite number of partial waves. However the constraints derived here can be used to ensure that the $f_\ell(s)$ for $0 \leqslant s \leqslant 1$ are at least consistent with crossing. In fact they have been widely used to limit the values of parameters appearing in model amplitudes.

In section 2 the $f_\ell(s)$ are shown to satisfy, for $o \leqslant s \leqslant 1$, an infinite set of constraints which are sufficient for them to be solutions of the moment problem. This result is used in section 3 to derive constraints on integrals of $f_o(s)$ over the above region by considering a two dimensional moment problem. Finally the conclusions of this work are summarised in section 4.

## 2. Partial Waves and the Moment Problem

A very simple constraint can be derived immediately from the dispersion relation which is obtained from (8) by the interchange of s and t. Projecting out the partial-waves from this new form for F(s,t,u), one gets for $\ell=2,4,6\ldots$.the Froissart-Gribov representation.

$$f_\ell(s) = \frac{1}{\pi(1-s)} \int_1^\infty Q_\ell\left(\frac{2s'}{1-s} -1\right) \text{ Im } F(s'+i\varepsilon,s)\; ds'. \tag{13}$$

For $o \leqslant s \leqslant 1$ and $s' \geqslant 1$, $Q_\ell\left(\frac{2s'}{1-s} -1\right) > o$ and $\text{Im } F(s'+i\varepsilon,s) \geqslant o$ from (12), so that (Martin, 1967)

$$f\ (s) \geqslant o \qquad 1 \geqslant s \geqslant o \qquad \ell=2,4,\ldots \tag{14}$$

The odd partial waves are identically zero from the invariance of F(s,t,u) under interchange of t and u.

In fact one can improve on the inequalities (14) and show that the $f_\ell(s)$ satisfy conditions which are sufficient for them to be solutions of a "Hausdorff Moment Problem", (Common, 1969; Yndurain, 1969). The trick is to notice that the Legendre functions of the second kind have the representation

$$Q_\ell(x) = \int_o^\infty \left[x + (x^2-1)^{\frac{1}{2}} \cosh\theta\right]^{-\ell-1} d\theta, \tag{15}$$

for $x > 1$ . Substituting in (13),

$$f_\ell(s)=\frac{1}{2\pi}\int_{x_o}^{\infty}\int_{o}^{\infty}|x+(x^2-1)^{\frac{1}{2}}\cosh\Theta|^{-\ell-1}\ \mathrm{Im}F|\tfrac{1}{2}(x+1)(1-s)+i\varepsilon,s|\,d\Theta dx \tag{16}$$

with $x_o = (1+s)/(1-s)$.

By direct substitution, it is straight forward to prove that the $f_\ell(s)$ satisfy the infinite set of conditions (Yndurain, 1969),

$$\begin{aligned}
&\delta^o f_\ell \equiv f_\ell \geqslant o\\
&\delta^1 f_\ell \equiv u_o^{-2}\, f_\ell - f_{\ell+2} \geqslant o \qquad (17)\\
&\ldots\ldots\ldots\ldots\ldots\\
&\delta^k f_\ell = \sum_{j=o}^{k} \binom{k}{j} (-1)^j u_o^{2(j-k)} f_{\ell+2j} \geqslant o\\
&\ldots\ldots\ldots\ldots\ldots
\end{aligned}$$

where $u_o = (x_o^2 - 1)^{\frac{1}{2}}$ and $\ell = 2,4,\ldots\ldots$

This infinite set of conditions is both necessary and sufficient for the $f_\ell$'s to have the representation

$$f_\ell(s) = \int_{u_o}^{\infty} z^{-\ell-1} d\phi(z,s),\ o \leqslant s \leqslant 1;\ \ell = 2,4,6,\ldots \tag{18}$$

where $\phi(z,s)$ is a bounded non-decreasing function of z (Shohat and Tamarkin, 1943). By a simple change of variable this becomes

$$f_\ell(s) = \int_{o}^{1} u^{\frac{\ell}{2}-2}\ d\,\kappa(u,s)\ ,\ o \leqslant s \leqslant 1;\quad \ell = 2,4,6\ldots. \tag{19}$$

where $\kappa$ has the same properties as $\phi$. For reasons which will become apparent later the upper limit of the integral in (19) has been taken as 1 and not $u_o^{-1}$. This is allowable since $u_o^{-1} < 1$ and one can choose $\kappa(u,s)$ to be constant in the interval $[1/u_o,1]$.

## 3. Constraints on Averages of the S-wave over the Unphysical Region

The above constraints are not as useful as they might be since they relate D and higher waves, while most effort has been devoted to constructing models for S-and D-waves alone. To get round this difficulty, one can use the fact that integrals of the S-wave over the unphysical region $o \leq s \leq 1$, can be related through the crossing relation (6) to similar integrals of higher waves, (Balachandran and Nuyts, 1968; Roskies, 1970a; Basdevant et. al., 1969)

In particular defining,

$$I_{\ell,n}= \int_0^1 ds(1-s)^{\ell+1} \; s^n \; f_\ell(s), \; n=o,1,\ldots; \quad \ell=2,4,\ldots. \tag{20}$$

then

$$I_{2,o}= 2 \int_0^1 ds(1-s) \; (2s^2-s) \; f_o(s) \geq o$$

$$I_{2,1}= \int_0^1 ds(1-s) \; (-5s^3+7s^2-2s) \;\; f_o(s) \geq o$$

$$I_{2,2}= \int_0^1 ds(1-s) \; (s^4-4s^3+4s^2-s) \;\; f_o(s) \geq o$$

$$I_{2,3} = \int_0^1 ds(1-s)(-7s^5+20s^4-25s^3+15s^2-3s)\ f_o(s) \geqslant 0$$

$$I_{4,0} = \int_0^1 ds(1-s)(18s^4-32s^3+18s^2-3s)\ f_o(s) \geqslant 0$$

$$I_{4,1} = \int_0^1 ds(1-s)(-21s^5+51s^4-44s^3+16s^2-2s)\ f_o(s) \geqslant 0$$

These are the only $I_{n,\ell}$ which can be written in terms of the s-wave alone (Pennington, 1970 and 1971).

Since $f_\ell(s) \geqslant 0$ for $0 \leqslant s \leqslant 1$, it follows that the $I_{\ell,n}$ are for fixed $\ell$ a solution of a Hausdorff moment problem, i.e. they have the representation

$$I_{\ell,n} = \int_0^1 s^n\, d\psi_\ell(s), \qquad \ell = 2,4,\ldots; \qquad n = 0,1,2,\ldots \tag{22}$$

where $\psi_\ell$ is a bounded non-decreasing function of s in the interval $|0,1|$.

Necessary and sufficient conditions for the above representation of $I_{\ell,n}$ are, c.f. equation (23),

$$\begin{aligned}
&\delta^o I_{\ell,n} \equiv I_{\ell,n} \geqslant 0 \\
&\delta^1 I_{\ell,n} \equiv I_{\ell,n} - I_{\ell,n+1} \geqslant o \\
&\cdots\cdots\cdots\cdots \\
&\delta^k I_{\ell,n} \equiv \sum_{j=0}^{k} \binom{k}{j} (-1)^j\, I_{\ell,n+j} \geqslant o \\
&\cdots\cdots\cdots\cdots
\end{aligned} \tag{23}$$

However since only a finite number of the $I_{\ell,n}$ can be written in terms of the s-wave, necessary and sufficient conditions for a finite number of the $I_{\ell,n}$ to have the representation (22) are required. This is called a "Reduced moment problem" and, as is discussed by Common and Pennington (1971a), the required conditions are not those obtained by taking the first few of (23) which are only necessary but not sufficient. They are in fact given by the following theorem (see for example Akheizer, 1965):-

Theorem 1. The necessary and sufficient condition for the set $\{I_{\ell,n};\ n=o,\ldots,N\}$ to have the representation (22) with $\psi_\ell(s)$ a bounded non-decreasing function of s is that for all polynomials

$$P_N(u) = \sum_{n=o}^{N} c_n u^n \tag{24}$$

which are non-negative for $o \leqslant u \leqslant 1$, then

$$\mu(P_N) \equiv \sum_{n=o}^{N} c_n I_{\ell,n} \geqslant o. \tag{25}$$

The necessity of the above condition is obvious since

$$\mu(P_N) = \int_o^1 P_N(u)\, d\psi_\ell(s), \tag{26}$$

but the sufficiency is much more difficult to prove.

The general form of a polynomial which is positive in the interval $[o,1]$ is (Akheizer, 1965),

N odd

$$P_N(u)=u\Big[\sum_{j=o}^{\frac{1}{2}(N-1)} a_j u^j\Big]^2 + (1-u)\Big[\sum_{j=o}^{\frac{1}{2}(N-1)} b_j u^j\Big]^2 \tag{27}$$

N even

$$P_N(u)= \Big[\sum_{j=o}^{N/2} e_j u^j\Big]^2 +u(1-u)\Big[\sum_{j=o}^{\frac{1}{2}(N-2)} d_j u^j\Big]^2, \tag{28}$$

where $\{a_j\}$, $\{b_j\}$, $\{e_j\}$ and $\{d_j\}$ are sets of arbitrary real numbers which are not all zero.

Theorem 1. may then be written in the form,

<u>Theorem 2.</u> The necessary and sufficient conditions for the set $\{I_{\ell,n};n=o,\ldots\ldots N\}$ to have the representation (21), with $\psi_\ell(s)$ as in Theorem 1, are that

$$N=2m+1 \quad \begin{cases} \sum_{j=o}^{m}\sum_{i=o}^{m} a_i a_j \; I_{\ell,i+j+1} \geq o \\ \sum_{j=o}^{m}\sum_{i=o}^{m} b_i b_j \; (I_{\ell,i+j} - I_{\ell,i+j+1}) \geq o \end{cases} \tag{29}$$

$$N=2m \quad \begin{cases} \sum_{i=o}^{m}\sum_{j=o}^{m} e_i e_j \; I_{\ell,i+j} \geq o \\ \sum_{i=o}^{m}\sum_{j=o}^{m} d_i d_j \; (I_{\ell,i+j+1} - I_{\ell,i+j+2}) \geq o \end{cases}$$

where $\{a_i\}$ etc. are as above.

By the theory of quadratic forms the conditions (29) are equivalent to the requirement that,

$$W_{oo}^{\;n} \geq o, \quad \begin{vmatrix} W^n_{oo} & W^n_{o1} \\ W^n_{1o} & W^n_{11} \end{vmatrix} \geq o, \ldots\ldots, \begin{vmatrix} W^n_{oo} & W^n_{o1} & \cdots & W^n_{om'} \\ \vdots & & & \vdots \\ W^n_{m'o} & \cdots & \cdots & W^n_{m'm'} \end{vmatrix} \geq o \tag{30}$$

for n=1,2,3,4; $W^1_{ij}$, $W^2_{ij}$, $W^3_{ij}$, $W^4_{ij}$ are respectively the coefficients of $a_i a_j$, $b_i b_j$, $c_i c_j$ and $d_i d_j$ in equations (29) and m'=m for n=1,2,3 and m'=m-1 for n=4.

These are the general conditions, but in the application of the above theorems considered here N=3. By inspection the general polynomial $P_3(u)$ which is positive for $o \leq u \leq 1$ is a linear combination with non-negative coefficients of :-

$$1,\ 1-u,\ (u-\alpha)^2,\ u(1-u),\ u(u-\alpha)^2,\ (1-u)(u-\beta)^2$$

where $\alpha$ and $\beta$ are arbitrary numbers.

These give the following conditions on the $I_{\ell,n}$ :-

$$P_N(u) = 1 \quad ; \quad I_{2,o} \geq o, \quad I_{4,o} \geq o$$

$$P_N(u) = 1-u \quad ; \quad I_{2,o}-I_{2,1} \geq o, \quad I_{4,o}-I_{4,1} \geq o$$

$$P_N(u) = (u-\alpha)^2 \quad ; \quad I_{2,o}\, I_{2,2} \geq I_{2,1}^{\ 2} \tag{31}$$

$$P_N(u) = u(1-u) \quad ; \quad I_{2,1} - I_{2,2} \geq o$$

$$P_N(u) = u(u-\alpha)^2 \quad ; \quad I_{2,1}\, I_{2,3} \geq I_{2,2}^{\ 2}$$

$$P_N(u) = (1-u)(u-\beta)^2 \quad ; \quad (I_{2,o}-I_{2,1})(I_{2,2}-I_{2,3}) \geq (I_{2,1}-I_{2,2})^2$$

These conditions do not constrain the over all magnitude of the integrals of the s-wave. This is to be expected since only the positivity of the $\mathrm{Im} f_\ell(s)$ has been used but not the fact that it is $\leq 1$. However they do constrain ratios of such integrals.

It is usual to consider the quantities

$$x = \frac{7}{2}\,\frac{\int_0^1 ds(1-s)(5s^3-3s^2)\ \ f_o(s)}{\int_0^1 ds(1-s)(2s^2-s)\ \ f_o(s)} \tag{32}$$

$$y = \frac{\int_0^1 ds(1-s)(18s^4-32s^3+18s^2-3s)\ \ f_o(s)}{\int_0^1 ds(1-s)(2s^2-s)\ \ f_o(s)}$$

The conditions (31) constrain x and y to lie in the region A B C D E shown in Figure 1, (Pennington, 1971).

To derive the above constraints, the positivity of $f_\ell(s)$ for $o \leqslant s \leqslant 1$ has been used.

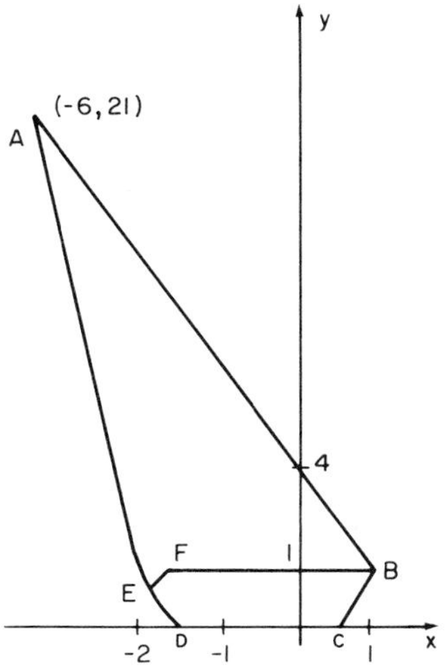


Figure 1. Allowed region for x and y from the positivity of ImF(s+iε,t).

One would expect to get tighter constraints by using the result that the $f_\ell(s)$ are solutions of a Hausdorff moment problem and this is in fact the case. Substituting from (19) into (20),

$$I_{\ell,n} = \int_0^1 \int_0^1 s^n \, y^{\ell/2 \, -1} \, d\eta(s,y) \tag{33}$$

where $y = (1-s)^2 u$ and $\eta(s,y)$ is a bounded non-decreasing function of both s and y.

The $I_{\ell,n}$ are solutions of a two dimensional Hausdorff moment problem and the constraints on $I_{\ell,n}$ are given by the following theorem, which is an obvious generalization of Theorem 1.

Theorem 3. Necessary and sufficient conditions for the set $\{I_{\ell,n};\ n=0,\ldots\ldots, N;\ \ell=2,\ldots\ldots, 2L+2\}$ to have the representation (33) is that for all real numbers $C_{jn}$, such that

$$P_{N,L}(u,s) \equiv \sum_{n=0}^{N} \sum_{j=0}^{L} C_{jn} \, u^j \, s^n \tag{34}$$

is non-negative when $0 \leqslant u \leqslant 1$, then

$$\sum_{n=0}^{N} \sum_{j=0}^{L} C_{jn} \, I_{2+2j,n} \geqslant 0. \tag{35}$$

As before only the cases when $0 \leqslant N \leqslant 3$ and $0 \leqslant L \leqslant 1$ have to be considered. It may be proved in that case by inspection that the most general form for $P_{N,L}(u,s)$ is a linear combination with + ve coefficients of the following forms :-

$1,\ s,\ (1-s),\ s(1-s),\ (\alpha-s)^2,\ s(\alpha-s)^2,\ (1-s)\,(\alpha-s)^2,$

$u, (1-u),\ us,\ (1-u)\,s,\ u(1-s),\ (1-u)(1-s),$

where $\alpha$ is an arbitrary constant. Each of these gives an independent constraint on the $I_{n,\ell}$'s and the complete set has been given by Common and Pennington (1971b).

This new set of conditions are much tighter as can be seen from Figure 1, the allowed values for x and y now constrained to lie in the region B C D E G.

4. Conclusions It was shown in the previous section that consequences of the $I_{\ell,n}$'s being a solution of a two-dimensional moment problem put strong constraints on $f_o(s)$. Various improvements can be made to these constraints to make them tighter, (see for example Griss, 1971), but they are quite technical and will not be considered here.

The unitarity condition used here is that $\text{Im}F(s+i\varepsilon,t) \geqslant o$ for $s \geqslant 1$ and $1 \geqslant t \geqslant o$, and this is obviously much weaker than requiring that $\text{Im}_\ell f\ (s) \geqslant o$ for all $\ell$. It is again technically difficult to take into account the latter conditions, but it has been done using methods which are essentially extensions of the moment problem technique, (Piguet and Wanders, 1969; Roskies, 1970b).

The tightest region obtained for the allowed values of x and y has been derived by Roskies and Yen (1971) and is the shaded area shown in Figure 2, where it compared with the previous area B C D E G. One can see that the constraint has been made appreciably tighter. In fact this most recent set of constraints on the $I_{\ell,n}$ is very tight and has been shown, by the above authors, to be violated by most models of the s-wave amplitude for $\pi^o\pi^o \to \pi^o\pi^o$ scattering. These authors have also shown that the allowed region for x and y cannot be made much tighter, by constructing model amplitudes which satisfy the positivity and crossing conditions and for which x and y almost fill the shaded area.

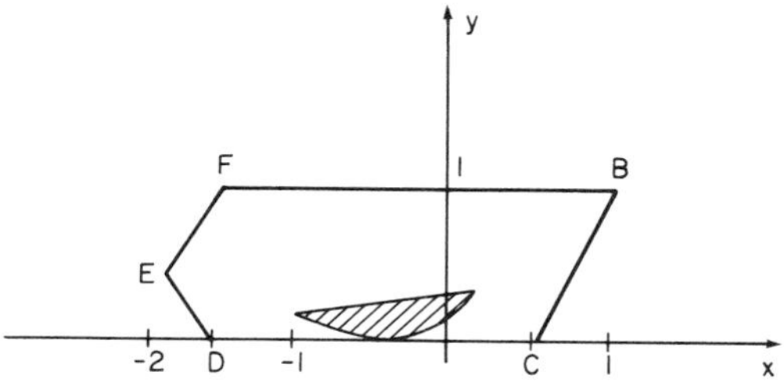


Figure 2. Allowed region for x and y from the positivity of the imaginary part of the Partial-wave Amplitudes.

The aim of this paper has been to introduce certain properties of solutions of moment problems and to show by example how useful they can prove to be in practice. In fact the requirement that a set of quantities is a solution of a moment problem puts very strong constraints on these quantities. Finally it should be stressed that when this set is finite, it is important to consider the reduced moment problem and the constraints given by Theorems 1 and 3. The reason for this is that the first few of the conditions (23) are necessary but not sufficient for the solution of this moment problem.

## REFERENCES

Akheizer N.I., (1965). "The Classical Moment Problem", pp 71-76, Oliver and Boyd.

Balachandran A.P. and Nuyts J., (1968). Phys. Rev. 172 1821.

Basdevant J.L., Cohen-Tannoudji G. and Morel A., (1969). Nuovo Cimento 64A, 585.

Common A.K., (1969). Nuovo Cimento 63A, 863.

Common A.K. and Pennington M.R., (1971a). Nuovo Cimento Letters 2, 429. and (1971b). Nuclear Physics B34, 253.

Griss M.L., (1971). Phys. Rev. D4, 2482 (1971).

Jin Y.S. and Martin A., (1964). Phys. Rev. 135, B1375.

Martin A., (1963). Phys. Rev. 129, 1432. and (1967) Nuovo Cimento 47A , 265.

Pennington M.R., (1970). Nuclear Physics B24, 317 and (1971). Nuclear Physics B25, 621.

Piguet O. and Wanders G., (1969). Physics Letters 30B , 418.

Roskies R., (1970a). Nuovo Cimento 65A , 467.

Roskies R., (1970b). J. Math. Phys., 11, 2913.

Roskies R. and Yen H.C., (1971) Phys. Rev. D4, 1873.

Shohat J.A. and Tamarkin J.D., (1943). "The Problem of Moments", p 9. American Mathematical Society.

Yndurain F.J., (1969). Nuovo Cimento, 64A, 225.

# THE MOMENT PROBLEM AND STABLE EXTRAPOLATIONS WITH AN APPLICATION TO FORWARD Kp DISPERSION RELATIONS

C. López
(Universidad Autónoma de Madrid, Spain)
and
F. J. Ynduráin
(CERN, Geneva, Switzerland and Universidad Autónoma de Madrid)

## 1. Introduction

This paper presents, in a rigorous way, a set of results on the problem of reconstructing the distributions $\sigma(x)$ and $\rho(x)$ from the moments

$$\mu_n = \int_0^1 dx\, \sigma(x)\, x^n , \quad n=0,1,\ldots\to\infty \tag{1}$$

(the so-called Hausdorff moment problem), or the quantities

$$\lambda_n = \frac{1}{\pi}\int_0^1 dx\, \frac{\rho(x)}{z_n - x} , \quad 1<z_0<z_1<z_2 \ldots \text{ and } z_n\to\infty \text{ as } n\to\infty , \tag{2}$$

(which may be called a Stieltjes-Hilbert problem). We use the interval $[0,1]$ for definiteness; the generalisation to any finite interval $[a,b]$ is trivial. Also obvious is the extension of (2) to other sets of z's provided they are outside $[0,1]$. At the same time we will discuss the reconstruction, from the same $\mu_n$'s or $\lambda_n$'s, of the functions given by

$$\psi(z) = \frac{1}{\pi}\int_0^1 dx\, \frac{\sigma(x)}{z-x} \tag{3}$$

and

$$\phi(z) = \frac{1}{\pi}\int_0^1 dx\, \frac{\rho(x)}{z-x} . \tag{4}$$

Note that, when $0 \leq x \leq 1$,

$$\text{Im}\ \psi(x) = -\sigma(x) \quad \text{and} \quad \text{Im}\ \phi(x) = -\rho(x). \tag{5}$$

Also

$$\phi(z_n) = \lambda_n \quad \text{and} \quad \frac{\pi}{n!}\left(\frac{d}{du}\right)^n \left(\frac{\psi(u^{-1})}{u}\right)\Bigg|_{u=o} = \mu_n\ . \tag{6}$$

A brief survey of the results, together with physical motivations, may be found elsewhere (Yndurain, 1971, 1972). An application to the physically interesting problem of Kp forward dispersion relations will be discussed in section 5 as a practical example.

Several methods for dealing with problems associated with equations (1) to (4) have been discussed in the literature (Cutkosky, 1968; Ciulli, 1969; Pisút, 1970). We present here a different treatment which uses techniques of the moment problem and Padé approximants (Basdevant, 1969, 1972; Baker, 1969, Yndurain, 1972). Thus, we construct rational functions $\phi_N(z)$ and $\psi_N(z)$, such that they approximate with arbitrary accuracy the functions $\phi(z)$ and $\psi(z)$ respectively in the whole plane. The sense in which this "approximation" is to be understood will be explained below. Such $\phi_N(z)$ and $\psi_N(z)$ are given as explicit functionals of the quantities $\lambda_o, \lambda_1, \ldots, \lambda_N$ and $\mu_o, \mu_1, \ldots, \mu_N$. For physical applications it is not only necessary that (in some sense of equality)

$$\left.\begin{aligned} \lim_{N\to\infty} \phi_N^{\{\lambda_o,\ldots,\lambda_N\}}(z) &= \phi(z) \\ \lim_{N\to\infty} \psi_N^{\{\mu_o,\ldots,\mu_N\}}(z) &= \psi(z) \end{aligned}\right\} , \tag{7}$$

but one also needs stability of the $\phi_N^{\{\lambda_o,\ldots,\lambda_N\}}$ or $\psi_N^{\{\mu_o,\ldots,\mu_N\}}$ against small perturbations ("experimental errors") of the input $\lambda_o,\ldots,\lambda_N$ or $\mu_o,\ldots,\mu_N$. Mathematically this means that, considered as functionals of the $\lambda_o,\ldots,\lambda_N$ or $\mu_o,\ldots,\mu_N$,

$\phi_N^{\{.\}}$ or $\psi_N^{\{.\}}$ should be continuous.

Loosely speaking, we are able to solve these problems in the following sense: if $\rho$ or $\sigma$ is positive, we are able to show that the diagonal Padé approximants $\phi_N(x)$ or $\psi_N(z)$ converge towards $\phi(z)$ or $\psi(z)$ respectively pointwise if $z \notin [0,1]$, and in the mean if $z \in [0,1]$. That is to say, when $f(z)$ belongs to a suitable class

$$\int_0^1 \phi_N f \to \int_0^1 \phi f, \quad \int_0^1 \psi_N f \to \int_0^1 \psi f \quad \text{as} \quad N \to \infty .$$

It should be obvious that one cannot get better than convergence in the mean because $\phi$ or $\psi$ are, in general, distributions when $z \in [0,1]$.

Note that by virtue of (5), this solves at the same time the problem of reconstructing $\rho$, $\sigma$, and $\phi(z)$, $\psi(z)$ for $z \notin [0,1]$. It is also important that not only can we prove convergence, but that the rate of convergence is explicitly given.

If $\rho$, $\sigma$ are not positive, we are still able, in the general case, to reduce the problem to another problem where the new $\tilde{\rho}$, $\tilde{\sigma}$ are positive, so that we can apply the methods found for positive $\rho$, $\sigma$. In the important case where one can write $\rho = \rho_1 - \rho_2$, $\sigma = \sigma_1 - \sigma_2$ with $\rho_i \geq 0$, $\sigma_i \geq 0$, we also give a direct discussion. In all cases, however, the reconstruction is shown to be stable against small perturbations.

## 2. Complete Mesh of Weight Functions

Suppose that we are given a double array of functions defined in the interval $[0,1]$, $f_m^{(n)}(x)$. We will say that they form a complete linear mesh of weight functions (CLMW) if the following conditions are satisfied:

L1) For every $n$, $m$, $f_m^{(n)}(x)$ has $\nu$ piecewise continuous derivatives. This will be denoted by writing

$$f_m^{(n)}(x) \in C^{\nu}(0,1) .$$

The value of $\nu$ will depend on the singularities of the object to be averaged. For example, if it is a measure, $\nu=1$ or if it is a distribution $\nu=\infty$.

L2) $f_m^{(n)}(x) \geqslant 0$, and $\int_0^1 dx\, f_m^{(n)}(x) = 1$ for any m,n. (8)

L3) Given any point $x_0 \in [0,1]$, there exists a subsequence $f_{m_k}^{(n_k)}(x)$ such that

$$\lim_{k\to\infty} f_{m_k}^{(n_k)}(x) = \delta(x-x_0) \tag{9}$$

and

$$\lim_{k\to\infty} f_{m_k}^{(n_k)}(x) = 0 \quad \text{if} \quad x \notin [x_0-\delta,\, x_0+\delta] \tag{10}$$

for all $\delta>0$; this last limit is required to be uniform.

These conditions just prescribe that, for fixed n, the $f_m^{(n)}$ form a set of weights which, as n increases, becomes sharper and stronger. We may define

$$\bar{x}_m^{(n)} \equiv \int_0^1 dx\, f_m^{(n)}(x)\, x \tag{11}$$

$$\delta_m^{(n)} \equiv \int_0^1 dx\, f_m^{(n)}(x) \left[x-\bar{x}_m^{(n)}\right]^2 \tag{12}$$

to be the point around which the average is taken and the precision of the average, respectively. Thus L3 just means that, in the limit, the $\{f_m^{(n)}\}$ permit infinitely precise averages.

We shall now discuss two specific CLMW's that will be very useful to us.

Theorem 2.1: The set

$$b_j^{(n)}(x) \equiv \frac{1}{N_j^{(n)}}\, x^j(1-x)^{n-j}, \quad n=0,1,\ldots; \quad j=n,n-1,\ldots,0 \tag{13a}$$

where

$$N_j^{(n)} \equiv \int_0^1 dx\, x^j(1-x)^{n-j} \tag{13b}$$

forms a CLMW.

The $b_j^{(n)}$ may be called Bernstein-Stieltjes polynomials.

Proof: L1,2 are obvious, so we turn to proving L3. Let $x_o \in [0,1]$ be given; select a sequence j,n such that $j/n \to x_o$.

It is easy to check that the maximum of $b_j^{(n)}(x)$ is located at $x=j/n$. Then we will see that

$$\lim_{\substack{n,j\to\infty \\ j/n\to x_o}} \frac{\int_o^{y_o} dx\, b_j^{(n)}(x)}{\int_o^1 dx\, b_j^{(n)}(x)} = 0 \quad \text{if} \quad y_o < x_o ,$$

$$\lim_{\substack{n,j\to\infty \\ j/n\to x_o}} \frac{\int_{y_o}^{1} dx\, b_j^{(n)}(x)}{\int_o^1 dx\, b_j^{(n)}(x)} = 0 \quad \text{if} \quad y_o > x_o .$$

For definiteness, we assume $x_o, y_o \neq 0,1$. The extension to the endpoints is left to the reader. Owing to the positivity of the b, this will be sufficient (Schwartz, 1950) to complete the proof of L3, hence of theorem 2.1.

Take, for example, the first case: $y_o < x_o$. Clearly,

$$\frac{\int_o^{y_o} dx\, b_j^{(n)}(x)}{\int_o^1 dx\, b_j^{(n)}(x)} < \frac{y_o}{\int_{x_o-\varepsilon}^{x_o+\varepsilon} dx \left(\frac{x}{y_o}\right)^j \left(\frac{1-x}{1-y_o}\right)^{n-j}}$$

for all $\varepsilon$, $\varepsilon>0$, $x_o - y_o > \varepsilon$. As $j/n \to x_o$, we may replace the last expression by

$$\frac{y_o}{\int_{x_o-\varepsilon}^{x_o+\varepsilon} dx \left[\left(\frac{x}{y_o}\right)^{x_o} \left(\frac{1-x}{1-y_o}\right)^{1-x_o}\right]^n} .$$

One can easily verify that

$$\left(\frac{x}{y_o}\right)^{x} \left(\frac{1-x}{1-y_o}\right)^{1-x} \geqslant \lambda > 1$$

for all $x \in [x_o-\varepsilon,\ x_o+\varepsilon]$, so that

$$\left[\left(\frac{x}{y_o}\right)^{x_o} \left(\frac{1-x}{1-y_o}\right)^{1-x_o}\right]^{n} > \lambda^{n} ,$$

from which the desired result follows directly.

It is clear that the average point and dispersion of the b's are given, respectively, by

$$\bar{x}_j^{(n)} = j/n , \qquad \delta_j^{(n)} = \int_0^1 dx\ b_j^{(n)}(x)\ [x-j/n]^2 .$$

A second example of CLMW is furnished by the following:

Theorem 2.2: Let us consider the sequence of reals, $z_o, z_1, \ldots,$ with

$$1 < z_o < z_1 < \ldots < z_n < \ldots \to \infty ; \quad \sum_o^\infty \frac{1}{z_n} < \infty . \tag{14}$$

Then, what may be called Bernstein-Szegö fractions,

$$\beta_j^{(n)}(x) = \frac{1}{M_j^{(n)}} \cdot \frac{x^j (1-x)^{n-j}}{\prod_{k=o}^{n} (1-x/z_k)} , \quad n = 0,1,\ldots ; \quad j = n, n-1, \ldots, 0, \tag{15a}$$

where

$$M_j^{(n)} = \int_0^1 dx\ \frac{x^j (1-x)^{n-j}}{\prod_{k=o}^{n} (1-x/z_k)} , \tag{15b}$$

form a CLMW.

Proof: Again here L1,2 are obvious. To prove L3 we just remark that owing to formulae (14),

$$\lim_{n\to\infty} \frac{1}{\prod_{k=0}^{n} \left(\frac{z_k - x}{z_k}\right)} = g(x) \ ,$$

with g analytic in a neighbourhood of $[0,1]$. The rest of the proof is then straightforward.

An important property of the $\beta$ is that they can be decomposed into simple fractions,

$$\left.\begin{aligned} \beta_j^{(n)}(x) &= \sum_{k=0}^{n} \frac{(-1)^{n-j+k}\ \gamma_k(n,j)}{z_k - x} \\ M_j^{(n)}\ \gamma_k(n,j) &= (-1)^{n-j+k} \prod_{k'=0}^{n} z_{k'} \frac{z_k^j(1-z_k)^{n-j}}{\prod_{k'\neq k} (z_{k'}-z_k)} \geqslant 0 \end{aligned}\right\} . \quad (16)$$

The average point for the $\beta$ is given by the explicit formula

$$\bar{x}_j^{(n)} = \sum_{k=0}^{n} (-1)^{n-j+k}\ \gamma_k(n,j) \left(z_k \log\frac{z_k}{z_k - 1} - 1\right) .$$

The dispersion is

$$\delta_j^{(n)} \equiv \int_0^1 dx\ \beta_j^{(n)}(x) \left[x - \bar{x}_j^{(n)}\right]^2 .$$

## 3. Reconstruction of $\phi,\psi,\rho,\sigma$ for $\rho,\ \sigma \geqslant 0$

The reconstruction of $\phi(z)$, $\psi(z)$ [as given by Eqs. (3) to (6)] for $z \in [0,1]$, and when $\rho$, $\sigma \geqslant 0$, is a classical problem whose solution can be found in the standard literature (Shohat, 1943). Incidentally, it will turn out that this reconstruction will also be recovered as a corollary of our discussion here; indeed, we will show how to recover averages

$$\int_0^1 dx\ \sigma(x)\ f(x) \ , \qquad \int_0^1 dx\ \rho(x)\ f(x) \ ,$$

when $f \in C^{\nu}(0,1)$, so that it will be sufficient to specialise in the particular case $f(x) = 1/(z-x)$. Thus, we will devote our efforts to the reconstruction of averages of $\rho(x)$, $\sigma(x)$, Re $\phi(x)$, Re $\psi(x)$ when $x \in [0,1]$.

The choice of the space of test functions is dictated by the singularities of $\rho, \phi, \ldots$ . In our case, the positivity of $\rho, \sigma$ makes it possible for us to choose a rather wide space of test functions, namely $C^1(0,1)$ as defined before. So, if we set up a sequence of approximating functions, say $\phi_N(x)$, we have two different problems. First, we have to show that $\phi_N(x) \to \phi(x)$ over $C^1(0,1)$. This, then, shows that convergence of averages with any CLMW holds and, therefore, we are free to reconstruct averages with any system of CLMW that is more convenient to us.

Before entering into the actual discussion, we will recall two standard results (Courant, 1953):

<u>Lemma 3.1</u>: (Weierstrass's theorem.) For any $\varepsilon>0$, and any $f(x) \in C^1(0,1)$, there exists one polynomial $p_\varepsilon(x)$ with

$$p_\varepsilon(x) \leqslant f(x) \leqslant p_\varepsilon(x) + \varepsilon \; . \qquad (17)$$

<u>Lemma 3.2</u>: (Szegö's theorem.) Given $f(x) \in C^1(0,1)$, and for any $\varepsilon>0$, there exists a $r_\varepsilon(x)$ such that $(1<z_0<z_1<\ldots)$,

$$r_\varepsilon(x) = \sum_{k=0}^{\ell} C_k/(z_k-x)$$

and

$$r_\varepsilon(x) \leqslant f(x) \leqslant r_\varepsilon(x) + \frac{\varepsilon}{z_0-x} \; . \qquad (18)$$

For the definition and properties of Padé approximants of the first or second kinds, we shall use the conventions of Yndurain (1972). Let

$$\Psi(u) = u^{-1}\, \psi(u^{-1}) \;, \quad \text{where} \quad u=z^{-1} \; .$$

Then

$$\Psi(u) = \frac{1}{\pi} \int_0^1 dx \, \frac{\sigma(x)}{1-ux} \quad .$$

Let us form the $[M/N]$ approximants (of the first kind) to $\Psi(u)$ and define $\psi_{N,M}(z)$ by

$$\psi_{N,M}(z) = \psi_{N,M}(u^{-1}) = u\{[M/N] \; \Psi(u)\}. \qquad (19)$$

If $\sigma$ is positive, we can prove the following properties of the $\psi_{N,M}$:

$$\left.\begin{aligned} \psi_{N,M}(z) &= \frac{1}{\pi} \int_0^1 dx \, \frac{\sigma_{N,M}(x)}{z-x} \\ \sigma_{N,M}(x) &= \sum_{k=0}^{M'} C_k \; \delta(x-x_k) \geq 0, \; x_k \in [0,1] \end{aligned}\right\} . \qquad (20)$$

It is also obvious from the very construction that

$$\mu_n = \int_0^1 dx \; \sigma_{N,M}(x) \; x^n \; , \quad \text{if} \quad n=0,1,\ldots,N+M. \qquad (21)$$

Actually, it is formula (21) that we shall use almost invariably. Indeed, most of our results are after all independent of the use of Padé or other types of approximants, provided only that such approximants verify formula (21). The Padé choice has as its main merit that of being simple and already available to us from extensive literature, but it is quite possible that one could find better choices for particular problems. Also, it is clear that in principle any choice of N,M would do; however, to reproduce correctly the asymptotic properties of $\psi(z)$ we shall always take the diagonal approximants with N=M. We shall thus simply write $\sigma_N$, $\psi_N$ for $\sigma_{N,N}$, $\psi_{N,N}$.

From formula (21) one obviously has

$$\int_0^1 dx \; \sigma_N(x) \; b_j^{(n)}(x) = \int_0^1 dx \; \sigma(x) \; b_j^{(n)}(x), \; n \leq 2N. \qquad (22)$$

Formula (21) also implies that for any f in $C^1(0,1)$,

$$\lim_{N\to\infty} \int_0^1 dx\ \sigma_N(x)\ f(x) = \int_0^1 dx\ \sigma(x)\ f(x)\ .$$

In fact, take any $\varepsilon>0$; due to Lemma 3.1 we can find $p_\varepsilon(x)$ satisfying formula (17). Let $\nu$ = degree of $p_\varepsilon$. Taking $2N\geq\nu$ we see that on the one hand, and since $\sigma_N\geq 0$ ,

$$\int_0^1 dx\ \sigma_N(x)\ p_\varepsilon(x) \leq \int_0^1 dx\ \sigma_N(x)\ f(x) \leq \int_0^1 dx\ \sigma_N(x)\ p_\varepsilon(x)\ +$$

$$+\ \varepsilon \int_0^1 dx\ \sigma_N(x) = \int_0^1 dx\ \sigma_N(x)\ p_\varepsilon(x) + \varepsilon\mu_0\ ,$$

and, on the other hand,

$$\int_0^1 dx\ \sigma(x)\ p_\varepsilon(x) \leq \int_0^1 dx\ \sigma(x)\ f(x) \leq \int_0^1 dx\ \sigma(x)\ p_\varepsilon(x) + \varepsilon\mu_0\ ;$$

since, owing to formula (21)

$$\int_0^1 dx\ \sigma(x)\ p_\varepsilon(x) = \int_0^1 dx\ \sigma_N(x)\ p_\varepsilon(x)\ ,$$

it follows that

$$\left|\int_0^1 dx\ \sigma(x)\ f(x) - \int_0^1 dx\ \sigma_N(x)\ f(x)\right| \leq \mu_0\varepsilon\ , \tag{23}$$

Q.E.D.

As a matter of fact, formula (23) not only proves convergence, but gives the explicit rate of convergence.

This result has two consequences: first, it shows that one has convergence for any CLMW, so we may choose whichever CLMW we like, as they are all equivalent. However, the most convenient one will, in general, be the $\{b_j^{(n)}\}$. So we have performed the desired reconstruction of $\sigma$ from the $\mu_n$. Secondly, and as noted in the Introduction and in the beginning of this section, also the reconstruction of $\psi(z)$, $z \notin [0,1]$ (now pointwise) is a consequence of formula (23).

It only remains to be proven that, now in the mean,

$$\mathrm{Re}\ \psi_N(x) \to \mathrm{Re}\ \psi(x)\ ,\quad x \in [0,1]\ .$$

To do so, one notes that we can write

$$\int_0^1 dx\ \mathrm{Re}\ \psi(x)\ f(x) = \int_0^1 dx\ \sigma(x)\ \tilde{f}(x)\ ,$$

$$\int_0^1 dx\ \mathrm{Re}\ \psi_N(x)\ f(x) = \int_0^1 dx\ \sigma_N(x)\ \tilde{f}(x)\ ,$$

where

$$\tilde{f}(x) \equiv \frac{1}{\pi}\int_0^1 dy\ \frac{1}{y-x}\ f(x)\ .$$

Now, it is known (Muskhelishvili, 1953) that $f \in C^1(0,1)$ implies that also $\tilde{f} \in C^1(0,1)$, so formula (23) still proves that

$$\lim_{N\to\infty}\int_0^1 dx\ \mathrm{Re}\ \psi_N(x)\ f(x) = \int_0^1 dx\ \mathrm{Re}\ \psi(x)\ f(x)$$

for any $f \in C^1(0,1)$. [We assume $f(0) = f(1) = 0$ to avoid end-point singularities.]

To solve the Stieltjes-Hilbert problem, we need so-called second-kind Padé approximants. With $\phi(z)$ defined by formulae (4), they are given by

$$\phi_N(z) = \cfrac{a_0/z}{1 + \cfrac{(1/z_0 - 1/z)a_1}{\ddots\ \cfrac{}{1 + (1/z_{N-1} - 1/z)a_N}}} \qquad (24a)$$

and the a's are fixed by requiring

$$\phi_N(z_n) = \lambda_n\ ,\quad n = 0,1,\ldots,N\ . \qquad (24b)$$

Alternative ways of obtaining the a's can be found in the literature. One may write

$$\phi_N(z) = \frac{p_{N'}(z)}{q_{N'+1}(z)}, \quad N' = \begin{cases} N/2 & \text{if } N \text{ even} \\ (N-1)/2 & \text{if } N \text{ odd} \end{cases} \tag{25}$$

where $p_{N'}$, $q_{N'+1}$ are polynomials of degrees $N'$, $N'+1$, and it can be shown that when $\rho$ is positive,

$$\rho_N(x) \equiv -\mathrm{Im}\ \phi_N(x) = \sum_{0}^{N'} b_n \delta(x-x_n) \geqslant 0,\ x_n \in [0,1]. \tag{26}$$

Clearly,

$$\int_0^1 dx\ \rho_N(x)\ \beta_k^{(n)}(x) = \int_0^1 dx\ \rho(x)\ \beta_k^{(n)}(x) \quad \text{if} \quad n \leqslant N. \tag{27}$$

Furthermore, using formula (18), the construction performed for the first-kind Padé approximants can be repeated for the second-kind ones with no changes, and thus one can extend controlled convergence $\rho_N(x) \to \rho(x)$ to $C^1(0,1)$. In particular, this also implies convergence of $\phi_N(z)$ to $\phi(z)$ pointwise if $z \notin [0,1]$, or in the mean if $z \in [0,1]$. So we are free to use any CLMW, as they are all equivalent; in particular, the $\{\beta_j^{(n)}\}$ will be convenient.

All this still leaves open the problem of optimisation of the estimates of averages, $\int dx\ \sigma f$, $\int dx\ \rho f$. A complete discussion of this can be found in Shohat and Tamarkin (1943) and Wheeler and Gordon (1970), and, in the particular case in which $f(x) = 1/\pi(x-z)$, in Baker (1969) and Common (1969).

In the preceding section we have constructed functionals:

$$\sigma_N^{\{\mu\}},\ \psi_N^{\{\mu\}}\ ;\quad \rho_N^{\{\lambda\}},\ \phi_N^{\{\lambda\}}\ ,$$

of

$$\mu_0, \mu_1, \ldots, \mu_{2N}\ ;\quad \lambda_0, \lambda_1, \ldots, \lambda_N$$

such that, in various senses,

$$\sigma_N^{\{\mu\}} \to \sigma,\ \psi_N^{\{\mu\}} \to \psi, \text{ etc.}$$

It will now be shown that these approximations are stable. To be precise, we will prove that, if we define the norm by:

$$\| \{\mu_o,\dots,\mu_{2N}\} - \{\mu'_o,\dots,\mu'_{2N}\} \| \equiv \sum_o^{2N} |\mu_k-\mu'_k| ,$$

$$\| \{\lambda_o,\dots,\lambda_N\} - \{\lambda'_o,\dots,\lambda'_N\} \| \equiv \sum_o^{N} |\lambda_k-\lambda'_k| ,$$

then

$$\lim_{\{\mu'\}\to\{\mu\}} \psi_N^{\{\mu'\}} = \psi_N^{\{\mu\}} ,$$

$$\lim_{\{\lambda'\}\to\{\lambda\}} \phi_N^{\{\lambda'\}} = \phi_N^{\{\lambda\}} .$$

Furthermore, we shall also be able to give the explicit rate of convergence. As all cases are similar, we shall prove this in some detail for $\sigma_N$. Also, it will be clear that the same conclusions will hold true for the exact $\sigma,\rho$.

Let $f \in C^1(0,1)$ be given. First of all, we use formula (17) to construct $p_\varepsilon(x) = \Sigma_o^\nu a_{\varepsilon k}x^k$. Then we take $2N\geq\nu$, and assume that

$$\| \{\mu'_o,\dots,\mu'_{2N}\} - \{\mu_o,\dots,\mu_{2N}\} \| < \delta .$$

Then, it is easy to check, using positivity, that

$$\left.\begin{aligned} \left|\int_o^1 dx \left[\sigma_N^{\{\mu'\}}-\sigma_N^{\{\mu\}}\right] f\right| &\leq \varepsilon \max\{\mu_o,\mu'_o\} + \delta \max|a_{\varepsilon k}| \\ \left|\int_o^1 dx \left[\sigma^{\{\mu'\}}-\sigma^{\{\mu\}}\right] f\right| &\leq \varepsilon \max\{\mu_o,\mu'_o\} + \delta \max|a_{\varepsilon k}| \end{aligned}\right\} . \qquad (28)$$

Formulae (28) are appropriate for proving stability. They also have the merit of being readily generalisable to the case where $\sigma$ is not positive, although it still is decomposable as $\sigma = \sigma_1-\sigma_2$, $\sigma_i\geq 0$ (cf. Section 4). However, they do not give the best estimate for the instability. This estimate is given by

$$\left|\int_0^1 dx \left[\sigma^{\{\mu'\}}-\sigma^{\{\mu\}}\right] f\right| \leq \varepsilon \max\{\mu_o,\mu'_o\} + \max\left|\sum_o^\nu a_{\varepsilon k}\left[\mu'_k-\mu_k\right]\right| , \qquad (29)$$

where both maxima are taken over all values of $\mu,\mu'$ compatible with the conditions, valid for all $k,\ell$ with $k+\ell\leq\nu$,

$$\sum_{0}^{\nu} |\mu_k'-\mu_k| < \delta, \ \Delta_{\ell k}\mu \geq 0, \ \Delta_{\ell k}\mu' \geq 0 , \qquad (30)$$

the $\Delta\mu$, $\Delta\mu'$ being defined in next section, Eq. (31a).

## 4. Oscillating $\rho,\sigma$

If one does not restrict $\rho$ or $\sigma$ by requiring them to be positive, the results one gets are not as clear-cut as before. One of the reasons is that the singularities of $\phi,\psi$ may be very violent. Then the methods used to prove that the zeros of the denominators of the Padé approximants lie in $[0,1]$ cannot be applied, so we have no control over the discontinuities of $\phi_N$, $\psi_N$. Nevertheless, answers can be given and we shall present them in this section. Before doing so, we shall recall a result that will be useful:

Lemma 4.1: A set of necessary and sufficient conditions for the $\mu_n$ to possess the representation (1) with $\sigma\geq 0$ is that the inequalities

$$\Delta_{jn}\mu \equiv \sum_{k=0}^{j} \binom{j}{k} (-1)^k \mu_{n+k} \geq 0 \qquad (31a)$$

be satisfied for all $n,j$. Similarly, in order for the $\lambda_n$ to satisfy formula (2) with $\rho\geq 0$ it is necessary and sufficient that

$$D_{jn}\lambda \equiv \sum_{k=0}^{n} (-1)^{n-j+k} \gamma_k(n,j)\, \lambda_k \geq 0 \qquad (31b)$$

for all $n,j$. For the proofs, see for example, Yndurain (1972). Conditions (31) are only complete if we know all the $\mu,\lambda$. If only a finite number is available, extra conditions have to be added (Shohat and Tamarkin, 1943).

Now we turn to the solution of the reconstruction problems. As before, we will consider in detail the Hausdorff case, indicating the modifications necessary to extend the discussion to the Stieltjes-Hilbert problem.

Theorem 4.1: If we can write $\sigma(x) = \sigma_1(x) - \sigma_2(x)$ with $\sigma_i \geq 0$, and one of the $\sigma_i$ (for example, $\sigma_2$) is bounded: $\sigma_2(x) \leq L$, $x \in [0,1]$, then one can reduce the discussion to the case treated in Section 3. It will be obvious that the same holds substituting $\sigma$ by $\rho$.

Proof: Define $\bar{\sigma}(x) \equiv \sigma(x) + L$. Then, $\bar{\sigma}(x) \geq 0$. Therefore, if we set

$$\bar{\mu}_n \equiv \int_0^1 dx\ \bar{\sigma}(x)\ x^n ,$$

we will be in the conditions of Section 3. It is then sufficient to note that

$$\bar{\mu}_n = \mu_n + \frac{L}{n} , \quad \bar{\psi}(z) = \psi(z) + \frac{L}{\pi} \log \frac{z}{z+1} \tag{32}$$

to recover approximants to $\psi$ from approximants to $\bar{\psi}$. Remark that from a practical point of view the existence of the bound L (and its computation) can be obtained by writing formula (32) and requiring that $\bar{\mu}_n$ verify formula (31a).

Theorem 4.2a: If there exists some $\nu$ such that the $\nu^{th}$ primitive of $\sigma$,

$$\sigma^{(-\nu}(x) \equiv \int_0^x dx_1 \int_0^{x_1} dx_2 \ldots \int_0^{x_{\nu-1}} \sigma(x_\nu) ,$$

is positive, we can still reduce the discussion to the case of Section 3.

Proof: Consider the case $\nu=1$; it will be clear that the general case can be reduced to this by iteration. Starting with the Hausdorff problem, we may write, by partial integration,

$$\mu_n = \int_0^1 dx\ \sigma(x)\ x^n = \mu_0 - n \int_0^1 dx\ \sigma^{(-1}(x)\ n^{n-1} .$$

We have assumed that $\sigma(x)$ is not singular at x=0, x=1. This is always possible, otherwise we would change the range of integration to $[-\delta, 1+\delta]$ and change variables to bring this back to $[0,1]$, so $\sigma$ would now vanish in neighbourhoods of 0 and 1. Hence, if we define

$$\mu_n^{(-1} \equiv \int_0^1 dx\, \sigma^{(-1}(x)\, x^n \,, \quad n=0,1,\ldots,$$

the discussion of Section 3 may be repeated changing $\sigma$ by $\sigma^{(-1}$. To recover from approximants to $\psi^{(-1}$ approximants to $\psi$, we just remark that

$$\mu_n^{(-1} = \frac{1}{n+1}(\mu_0 - \mu_{n+1}) \,, \qquad \psi(z) = \frac{\mu_0}{\pi(z-1)} + \frac{d}{dz}\psi^{(-1}(z). \tag{33}$$

Just as before, the existence of $\nu$ with $\sigma^{(-\nu} \geqslant 0$ can be obtained, and the value of $\nu$ calculated, by requiring the $\mu_n^{(-\nu}$ to satisfy formula (31a).

For the Stieltjes-Hilbert problem the generalisation is slightly complicated. Here we have:

<u>Theorem 4.2b</u>: If there exists some $\nu$ such that the $\nu^{th}$ "modulated" primitive of $\rho$,

$$\rho^{(\nu)}(x) \equiv \int_0^x dx_1 \frac{1}{(z_{\nu-1}-x_1)^\nu} \int_0^{x_1} dx \frac{1}{(z_{\nu-2}-x_2)^{\nu-1}} \cdots \int_0^{x_{\nu-1}} dx_\nu \frac{1}{z_0 - x}\, \rho(x_\nu) \,,$$

is positive, then we can reduce the discussion to the case of Section 3.

<u>Proof</u>: As before, we take $\nu$=1. Then,

$$\lambda_n \equiv \frac{1}{\pi}\int_0^1 dx \frac{\rho(x)}{z_0 - x} \cdot \frac{z_0 - x}{z_n - x} = \frac{z_0 - 1}{z_n - 1}\lambda_0 + \frac{z_n - z_0}{\pi}\int_0^1 dx \frac{\rho^{(1)}(x)}{(z_n - x)^2} \,.$$

We may thus define

$$\lambda_n^{(1)} \equiv \frac{1}{\pi}\int_0^1 dx\, \frac{\rho^{(1)}(x)}{(z_n-x)^2}\,. \tag{34}$$

Clearly, the relation between the $\lambda$'s and the $\lambda^{(\nu)}$'s is

$$\lambda_n^{(1)} = \frac{1}{z_n-z_0}\left\{\lambda_n - \frac{z_0-1}{z_n-1}\lambda_0\right\},$$

and between

$$\phi^{(1)}(z) = \frac{1}{\pi}\int_0^1 dx\, \frac{\rho^{(1)}(x)}{(z-x)^2}$$

and $\phi(z)$ as given by formula (4):

$$\phi(z) = \left\{\phi^{(1)}(z) - \frac{\lambda_0}{1-z}\right\}(z_0-z) + \lambda_0\,.$$

It only remains to extend the analysis of Section 3 to the case where $\lambda_n^{(1)}$ is given by formula (4) rather than by formula (2). For this, all we have to do is substitute the Bernstein fractions $\beta_j^{(n)}(x)$ by the fractions

$$\begin{aligned} \beta_{2,j}^{(n)}(x) &\equiv \frac{1}{M_{2,j}^{(n)}}\cdot\frac{\left[x^j(1-x)^{n-j}\right]^2}{\prod_{k=0}^{n}(1-x/z_k)} \\ M_{2,j}^{(n)} &\equiv \int_0^1 dx\, \frac{\left[x^j(1-x)^{n-j}\right]^2}{\prod_{k=0}^{n}(1-x/z_k)} \end{aligned} \tag{35}$$

with

$$\beta_{2,j}^{(n)}(x) = \sum_k \gamma_{2,k}(n,j)(z_k-x)^{-2}\,.$$

The straightforward verification that the $\beta_{2,j}^{(n)}$ possess the exact analogues to the properties of the $\beta_j^{(n)}$ is left to the reader.

It is still possible to treat the general case by using a combination of the techniques developed already in the section. Indeed, concentrating on the Hausdorff problem for definiteness, we consider formula (1) when $\sigma$ is any arbitrary distribution $\sigma \in$ '(0,1). It is then known (Schwartz, 1950) that in this case there always exists an integer $\nu$ such that the $\nu^{th}$ primitive, $\sigma^{(-\nu}(x)$, is continuous in $[0,1]$. Since $[0,1]$ is compact, this implies that $\sigma^{(-\nu}(x)$ is bounded: $|\sigma^{(-\nu}(x)| \leqslant L$. Therefore, the function $\bar{\sigma}^{(-\nu}(x) \equiv \sigma^{(-\nu}(x) + L$ is positive, so we can apply the techniques of Section 3 to the problem

$$\bar{\mu}_n^{(-\nu} = \int_0^1 dx\ \bar{\sigma}^{(-\nu}(x)\ x^n .$$

The relation between the $\bar{\mu}_n^{(-\nu}$, $\bar{\psi}^{(-\nu}(z)$ and $\mu_n$, $\psi(z)$ are found directly by combining Eqs. (32) and (33). The determination of $\nu$, L can be made by requiring the $\bar{\mu}_n^{(-\nu}$ to satisfy Eqs. (31a).

Although this solves the problem in principle, it should be clear that for practical purposes the situation is not so favourable. Indeed, every time we take a primitive, one can use one moment less: hence, and since the number of moments one has available in realistic situations is finite, we are losing information. It is for this reason that we shall give a reconstruction procedure which works directly for oscillating $\sigma(\rho)$, provided they have nothing worse than $\delta$-function singularities. As is known, this is equivalent to requiring that one may write $\sigma = \sigma_1 - \sigma_2$ ($\rho = \rho_1 - \rho_2$) with $\sigma_i(\rho_i)$ positive. We shall then show that from the $\mu_n(\phi_n)$ we shall be able to reconstruct $\psi(z)$ $[\phi(z)]$, pointwise for $z \notin [0,1]$, in the mean when $z \in [0,1]$. Whether one should use the previous methods, or the ones to be described, depends essentially on whether we can or cannot afford to lose one moment.

From the discussion in Sections 1 and 3, it should be clear that all we have to do is find bounds to quantities of the type $\int \sigma f$, $\int \rho f$, $f \in C^1(0,1)$.

Let us start with the Hausdorff problem. Given $f \in C^1(0,1)$, and given $\varepsilon>0$, we construct $p_\varepsilon(x)$ as in Eq. (17). Then it is straightforward to check that, if $p_\varepsilon = \Sigma\, a_{\varepsilon k}\, x^k$,

$$\sum_0^N a_{\varepsilon k}\, \mu_k + \varepsilon \int_0^1 dx\, \sigma_1(x) \leqslant \int_0^1 dx\, \sigma(x)\, f(x)$$

$$\geqslant \sum_0^N a_{\varepsilon k}\, \mu_k - \varepsilon \int_0^1 dx\, \sigma_2(x). \quad (36)$$

This provides the required upper and lower bounds which depend on the quantities $\int_0^1 dx\, \sigma_i(x)$; these can be estimated with the method of Baker and Gammel (1971).

The techniques for the Stieltjes-Hilbert problem are totally similar. Instead of Eq. (17) we may use Eq. (18) to get, if $r_\varepsilon(x) = \Sigma_0^N\, C_{\varepsilon k}/(z_k - x)$, the bounds

$$\sum_0^N C_{\varepsilon k}\, \lambda_k + \varepsilon \int_0^1 dx\, \frac{\rho_1(x)}{z_0 + x} > \int_0^1 dx\, \rho(x)\, f(x)$$

$$\geqslant \sum_0^N C_{\varepsilon k}\, \lambda_k - \varepsilon \int_0^1 dx\, \frac{\rho_2(x)}{z_0 - x}. \quad (37)$$

Finally, it should be clear that in the case when $\rho$ and $\sigma$ satisfy the conditions of Theorem 4.1, the discussion of stability can be translated with minor changes, showing that the reconstruction is still stable.

To finish with this section, we recall that, although we have proved reconstruction, stability, etc., for (say) $\bar{\phi}^{(-\nu}$, $\bar{\psi}^{(\nu}$, these properties get translated without change for $\phi,\psi$, because the operation of differentiation is continuous in the sense of distributions. Also the explicit bounds can be readily translated from $\bar{\phi}^{(-\nu}$, etc., to $\phi$, etc., by partially integrating back. For example, if we want to compute bounds for $\int_0^1 dx\, \rho(x)\, f(x)$, all we have to do is to compute bounds to $\int_0^1 dx\, \rho^{(-1}(x)\, f'(x)$ which just equals $\mu_0\, f(1) - \int_0^1 dx\, \rho(x)\, f(x)$.

## 5. Extrapolation in forward Kp dispersion relations

Forward Kp dispersion relations can be written as

$$f(E) = \int_{E_o}^{E_1} dE' \, \frac{\rho(E')}{E'+E} + \frac{X_\Lambda}{(E_\Lambda^2-\mu^2)(E+E_\Lambda)} \bar{G}^2 \,, \qquad (38a)$$

where

$$X_Y \equiv \frac{(M_Y-M)^2-\mu^2}{4M^2} \,, \qquad E_Y \equiv \frac{M_Y-M^2-\mu^2}{2M} \,,$$

$M_Y$ being the mass of hyperon Y, Y=Λ or Σ, and M,μ the masses of p,K. $\bar{G}^2 \equiv G_\Lambda^2 + [X_\Sigma(\mu^2-E_\Lambda^2)(E+E_\Lambda)/X_\Lambda(\mu^2-E_\Sigma^2)(E+E_\Sigma)] G_\Sigma^2 \simeq G_\Lambda^2$ , with $G_Y$ the KpY coupling constant. In terms of the real part of the scattering amplitudes for $K^\pm p$, $D_\pm(E)$, and of the total $K^\pm p$ cross-sections, $\sigma_\pm(E)$, f(E) is given by

$$f(E) = \frac{\pi}{E^2-\mu^2} \left[D_+(E) - \frac{1}{2} D_+(\mu)\left(1 - \frac{E}{\mu}\right) - \frac{1}{2} D(\mu)\left(1 + \frac{E}{\mu}\right)\right]$$

$$- \frac{1}{2\pi}\int_0^\infty dk' \, \frac{\sigma_+(k')}{E'(E'-E)} - \frac{1}{2\pi}\int_{k_1}^\infty dk' \, \frac{\sigma_-(k')}{E'(E'-E)} \,.$$

Here, $k^2 \equiv E^2-\mu^2$; $D_\pm, \sigma_\pm$ are experimentally accessible quantities, so one can know the values of f,

$$\lambda_n = f(E_n) \,, \quad n=0,1,\ldots \qquad (38b)$$

at different energies $E_n$. The quantity is related to the absorptive part for $K^-p$ scattering, $A_-$, by $\rho(E') = A_-(E')/(\mu^2-E'^2)$. $E_o = (m_\pi+\mu_\Lambda)^2$ is the beginning of the unphysical cut, and we take $E_1 = 0.49016$ GeV slightly below the opening of the physical cut, for technical reasons. The mathematical problem is contained in equations 38a and b. The problem here is typically one of extrapolation: given Eq. (38b), with some "experimental" errors

$$\lambda'_n \leqslant \lambda_n \leqslant \lambda''_n \qquad (39)$$

extrapolate (38a) to find $\bar{G}^2$ and averages of ρ. Further details of the strong interaction aspects of the problem are in Lopez thesis (1972). From physical reasons, one knows that it is possible to write $\rho(E') = \rho_1(E') - \rho_2(E')$, $\rho_i > 0$. However, it is possible to give good reasons for believing that $\rho_2 \ll \rho_1$, and, indeed, the experimental $\lambda_n$ are compatible with this through Eq. (31b). Therefore, we assume $\rho(E') \geq 0$. Other (numerical) examples of extrapolations with oscillating $\rho$'s have been worked out by Garcia and Paramio (1972), with excellent results.

Apart from the change of interval, $[0,1] \to \{E_\Lambda\} \cup [E_0,E_1]$, the methods of Section 3 can be applied straightforwardly. There are only a few points that have to be noted.

i) Since the point $E_\Lambda$ is disjoint from $[E_0,E_1]$, averages around $E'=E_\Lambda$ will just give the value and absolute bounds of $\bar{G}^2$.

ii) Applying the formulas (29) it turns out to be possible to go up to five averages of $\rho(E')$. This would permit us to calculate the third-order Padé approximant, which will approximate locally $f(E)$ for $E$ outside the cut. However, the first-order Padé approximant gives such a good approximation that, for $E$ outside the cut, it is pointless to go to higher orders.

iii) The higher-order averages of $\rho$ begin to grow instabilities, i.e. large absolute errors, computed from (29). It is, however, possible to construct higher averages if we substitute for (29) a formula giving statistical errors, where we choose to average over the $\beta$ as discussed by Lopez (1972b).

The explicit results are summarised in the following two figures. One should compare, for example, the estimates of $\bar{G}^2/4\pi$ with the results of the polynomial extrapolation of Cutkosky and Deo (1968), $\bar{G}^2/4\pi = 15^{+6}_{-4}$, and the positivity bounds of Rogers (1969), $\bar{G}^2/4\pi > 10$.

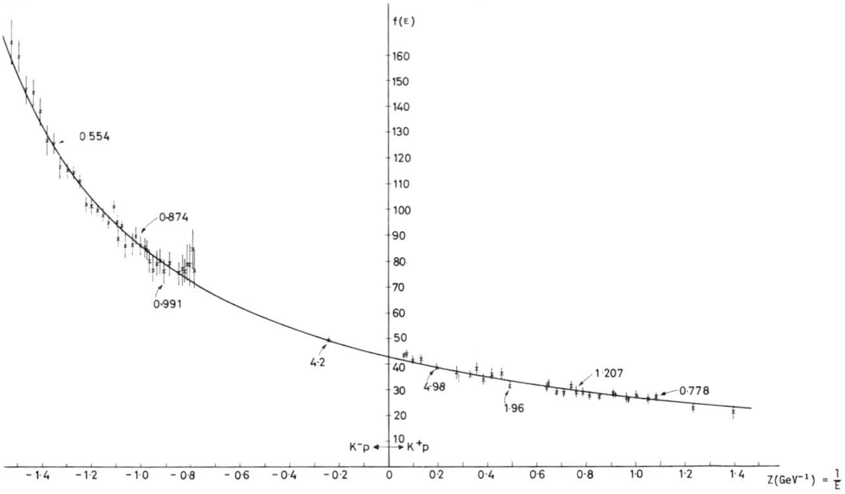


Fig. 1 Extrapolation to f(E) (outside the cut) with f(E) = pole + $a_0/(E+b_1)$, $a_0$ = 53.9 ± 2.7, $b_1$ = 0.45 ± 0.01. $\chi^2$ = 54.2/73.

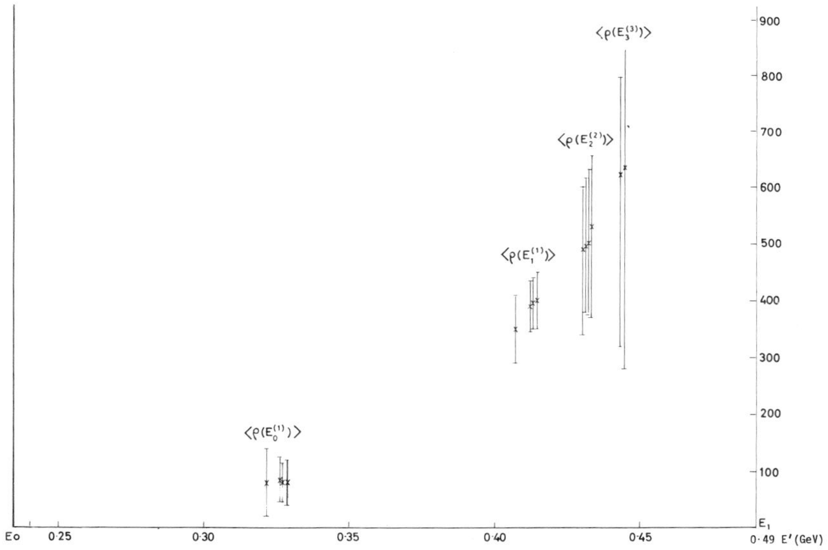


Fig. 2 Averages of ρ (inside the cut) with absolute errors. Values of $\bar{G}^2$: $\bar{G}^2/4\pi$ = 14.15 ± 1.17 (statistical); 12.39 ≤ $\bar{G}^2/4\pi$ ≤ 16.45 (absolute error bars).

References

Akhiezer N. I., (1943). The classical moment problem, Oliver and Boyd.

Baker G. A. Jnr., (1969). J. Math. Phys. 10, 814.

Baker G. A. Jnr. and Gammel J. L., (1971). J. Math. Anal. Applic. 33, 197.

Basdevant J. L., Bessis D. and Zinn-Justin J., (1969). Nuovo Cimento 64A, 185.

Basdevant J. L., (1971). Fortschr. Phys., to be published.

Ciulli S., (1969). Nuovo Cimento 61A, 787 and 62A, 301.

Common A. K., (1969). Nuovo Cimento 65A, 581.

Courant R. and Hilbert D., (1953). Methods of mathematical physics, (Interscience, New York), Vol. I.

Cutkosky R. and Deo B. B., (1968). Phys. Rev. 174, 1859.

Garcia A. and Paramio L., (1972), unpublished.

López C. and Yndurain F. J., (1972a). Phys. Letters, to be published.

López C., (1972b). Thesis, Universidad Autonoma de Madrid, Spain.

López C. and Yndurain F. J., to be published.

Muskhelishvili N. I., (1953). Singular integral equations (Nordhoof, Groningen).

Pisút J., (1970) in "Low-energy hadron interactions" (Springer Verlag, Berlin).

Rogers T. W., (1969). Phys. Rev. 178, 2478.

Schwartz L., (1950). Theorie des distributions (Hermann, Paris), Vol. I.

Shohat J. A. and Tamarkin J. D., (1943). The problem of moments, Amer. Math. Soc. Monographs, Providence.

Wheeler J. C. and Gordon R. G., (1970) in "The Padé approximant in theoretical physics" (Academic Press, New York).

Yndurain F. J., (1971). Ann. Phys. (N.Y.), to be published.

Yndurain F. J., (1972). The moment problem and applications, Lectures given at the 1972 Canterbury School on Padé Approximants and their Applications.

APPLICATION OF PADE APPROXIMANTS IN THE THREE-BODY PROBLEM

J. A. Tjon

(Institute for Theoretical Physics, University of Utrecht, The Netherlands)

## 1. Introduction

In recent years much progress has been made in the actual solution of the three-body problem. In particular, the three-nucleon system has been studied extensively (McKee et al., 1970). However, in most of the calculations up to now separable two-body interactions have been used, mainly because of the considerable simplifications they lead to with respect to the three-body equations (Faddeev, 1961). In most actual physical situations however we will be interested in more general types of two-body interactions. For example, the two-nucleon interaction is expected to be, at least partially, of a local type. For this reason several attempts have been made recently to study also the case of local interactions.

One particular line of approach has been followed by us in Utrecht and I would like to discuss some results we have obtained so far. The methods we have employed consist of the application of Padé techniques to the multiple scattering series of the Faddeev equations.

## 2. The Faddeev equations

For definiteness, let us consider a system of three identical scalar bosons of mass m which are interacting through two-body forces. Furthermore, let us suppose that two particles can be bound together and in addition, that there is only one boundstate possible. Restricting ourselves to the case of total angular momentum L=0, while the two-body potential can be approximated by only its s-wave part, the S-matrix element for describing the elastic scattering of a free particle with energy $q_i^2$ on a boundstate, characterised by the quantum numbers $\alpha$ is given by (Tjon, 1970)

$$M = 2\ {}_1\langle q_i,\alpha|V_2|q_i,\alpha\rangle_2 + \frac{2}{\sqrt{3}\,q_i}\int_0^\infty q\,dq \int_{A(q_i,q)}^{B(q_i,q)} p\,dp\ \psi_b(\sqrt{p^2+q^2-q_i^2}\,)U(p,q) \tag{1}$$

where

$$A(q,q') = |2q - q'|/\sqrt{3}$$
$$B(q,q') = |2q + q'|/\sqrt{3}$$

and

$$U(p,q) = \sum_{\substack{k,\ell,m \\ k\neq m,\ell\neq 1}} {}_\ell\langle p,q|T_k^\ell(s+i\varepsilon)|q_i,\alpha\rangle_m \tag{2}$$

with

$$T_k^\ell(s) = V_k\,\delta_{k\ell} - V_\ell\,G(s)\,V_k\ . \tag{3}$$

In the above equations $V_i$ represents the potential between particles j and k ($\neq$i), G(s) is the total Green's function $(H-s)^{-1}$ and $\psi_b$ is the two-particle boundstate wave function. Furthermore, $|p,q\rangle_1$ for example is the partial wave state of $|\vec{p},\vec{q}\rangle_1$ where $\vec{p}$ describes in the three-particle c-m system the relative momentum between particles 2 and 3, while $\vec{q}$ is related to the momentum of particle 1. We have

$$\vec{p} = \frac{1}{2\sqrt{m}} (\vec{k}_2 - \vec{k}_3)$$

$$\vec{q} = -\frac{\sqrt{3}}{2\sqrt{m}} \vec{k}_1$$

where $\vec{k}_i$ is the momentum of the $i^{th}$ particle. It is by now well known, that eq. (3) satisfies the Faddeev equation. In this case it takes the form

$$U(p,q) = \phi(p,q) - \frac{16\pi}{9\sqrt{3}} \int_0^\infty q'dq' \int_{A(q,q')}^{B(q,q')} p'dp' \frac{t_0(p,\bar{p}(q');\, s-q^2 + i\varepsilon)}{p'^2 + q'^2 - s - i\varepsilon} U(p',q') \qquad (4)$$

with

$$\bar{p}(q')^2 = p'^2 + q'^2 - q^2$$

and

$$\phi(p,q) = \frac{32\pi}{\sqrt{3}\, qq_i} \int_{A(q,q_i)}^{B(q,q_i)} p'dp' \, t_0(p,\bar{p}(q_i);\, s - q^2 + i\varepsilon)\, \psi_b(p'). \qquad (5)$$

In eqs. (4-5), $t_0$ is the off-shell two-particle T matrix for $\ell=0$, which satisfies the Lippmann-Schwinger equation.

From eq. (4) we see that we are dealing with an integral equation in two continuous variables. In the case of separable interactions this reduces to a one-dimensional integral equation. Due to this simplification the resulting equation can be solved by reducing it to quadratures and subsequently employing matrix inversion. However, for general two-particle interactions eq. (4) is extremely difficult to solve on present day computers with the above straightforward method. The approach we have followed in solving eq. (4) was to extract information from its Neumann solution with the aid of the ratio method in the three-particle boundstate region and the Padé method in the scattering region.

## 3. The three-particle boundstate

The Faddeev equation (4) may be written formally as

$$U(s,\lambda) = \phi(s) - \lambda\, K(s)\, U(s,\lambda) \qquad (6)$$

where we have introduced for convenience a complex parameter $\lambda$ in front of the kernel $K(s)$. Eq. (5) gives rise to the multiple scattering series for $M(s,\lambda)$ defined by eq. (1)

$$M(s,\lambda) = \sum_n \lambda^n M_n(s) \; . \qquad (7)$$

Here $M_o$ corresponds to the Born term, while $M_n$ represents the term which contains n times the two-particle T matrix.

In view of the compactness property of the kernel in eq. (4) (Faddeev, 1965) we know that the solution $U(s,\lambda)$ is a meromorphic function in $\lambda$ with possible poles at $\lambda_\alpha(s)$. Hence $U(s,\lambda)$ can be represented by

$$U(s,\lambda) = \sum_\alpha \frac{B_\alpha(s)}{\lambda_\alpha(s) - \lambda} - R(s,\lambda) \qquad (8)$$

where $R(s,\lambda)$ is an entire function of $\lambda$. Assuming that $\lambda_o(s)$ corresponds to the pole closest to the origin and that it is simple, one readily can show from eq. (8) that (Malfliet et al., 1969a)

$$\lim_{n\to\infty} \frac{U_{n+1}(s)}{U_n(s)} = \frac{1}{\lambda_o(s)} \; . \qquad (9)$$

In eq. (9) $U_n$ are defined as the coefficients of the perturbation expansion of U

$$U(s,\lambda) = \sum_n \lambda^n U_n(s) . \tag{10}$$

Since the groundstate energy of three particles with binding energy $s_o$ corresponds to a pole in $U(s,\lambda)$ at $s=s_o$ for the physical value $\lambda=1$, we see from eqs. (8-9) that the value of $s_o$ can in principle be determined from the condition that the ratios $r_n(s_o) \equiv U_{n+1}(s_o)/U_n(s_o)$ tend to 1 for $n\to\infty$. As a direct byproduct the corresponding boundstate wave function can also be determined from this calculation once $s_o$ has been computed. (Malfliet et al., 1969b). This can be seen in the following way. Let us consider the three-particle T matrix given by

$$T(s) = V - V\,G(s)\,V \tag{11}$$

with

$$V = V_1 + V_2 + V_3 . \tag{12}$$

Then according to the spectral decomposition theorem of G(s) we have

$$\lim_{s\to s_o} (s-s_o)\,{}_1\langle p,q|T(s)|p',q'\rangle_1 = \mathrm{const}(p^2+q^2-s_o)\,{}_1\langle p,q|\psi_t\rangle . \tag{13}$$

Here $\psi_t$ is the three-particle boundstate wavefunction and only the explicit dependence on the p and q variables are exhibited. Near the pole $s_o$ we may write instead of eq. (8)

$$U(s,\lambda) \simeq \frac{B_o(s_o)}{(s-s_o)\left(\dfrac{d\lambda_o(s)}{ds}\right)_{s=s_o}} \tag{14}$$

where it should be remembered that $B_o$ depends also on the p and q variables. Since there is a direct relation between $U(s,\lambda)$ and a matrix element of T(s) we may identify the residue $B_o(s_o)$ with the expression (13). Moreover, it can easily be seen from eq. (8) that

$$\lim_{n\to\infty} U_n(s_o) = B_o(s_o) \ . \tag{15}$$

Hence having determined $s_o$, the groundstate wave function can directly be computed from eqs. (13-15).

## 4. The scattering states

We now turn to discussing the scattering region, where the total three-particle energy s satisfies

$$s = s_B + q_i^2 \geq s_B \ . \tag{16}$$

In eq. (16) $s_B$ is the boundstate energy of the two-particle system. In the scattering case our task is to reconstruct the scattering amplitude $M(s,\lambda)$ from the multiple scattering series (7) at the physical point $\lambda=1$. Since we know that the solution $U(s,\lambda)$ is meromorphic in $\lambda$, we may use the property, that the Padé approximants $M_{N,N+J}(\lambda)=P_{N+J}(\lambda)/Q_N(\lambda)$ converge for $N\to\infty$ for all $\lambda$ except for certain points of measure zero (Chisholm, 1963, Nuttall, 1970 and Baker, 1972) and that

$$\lim_{N\to\infty} M_{N,N+J}(\lambda) = M(s,\lambda) \ . \tag{17}$$

One of the major problems in this approach is the determination of the coefficients $U_N(s)$. Furthermore, once these coefficients have been computed the sequences (9) and (17) should converge fast enough in order that the above described methods are practical. In the computation of $U_N(s)$ in the boundstate region no particular difficulties arise. They were found numerically from an iterative procedure of eq. (4). In the scattering region, however, we are dealing with integrations over functions having singularities. Below the inelastic threshold ($0<q^2<-s_B$) the only singularity arises from the two-particle boundstate pole in $t_o$. This singularity is removed by making a subtraction in the integral. Above the inelastic threshold additional singularities appear. In Fig. 1 the positions of the singularities are shown in the $p'$-$q'$ plane. For a given value of $q$ the integration region is between the lines $A(q,q')$ and $B(q,q')$. The singularities are located on the dashed parts of the quarter circle $p'^2+q'^2=s$ and of the straight line $q'=q_i$. The above singularities are all removed by making a number of subtractions. For the details we refer to Kloet and Tjon (1972). An additional complication in the inelastic region is that we also need to know the two-particle T-matrix for positive energy values. Since this cannot be computed by direct matrix inversion, we have followed the method of Kowalski (1965) and Noyes (1965). After carrying out the subtraction procedure, the resulting integrals were converted to Gaussian quadratures. Reasonably accurate results were obtained for the coefficients $M_n$ by using typically 12 and 20 meshpoints in the p and q variables respectively.

Fig. 1

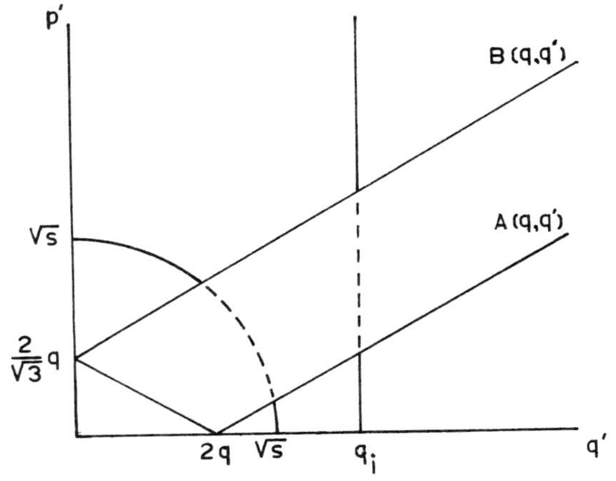

## 5. An example

As an illustrative example we studied the case of a Yukawa potential given by

$$V(r) = -\lambda_A \frac{e^{-\mu r}}{r} \quad . \tag{18}$$

Since we were particularly interested in the three-nucleon problem we have taken for the mass m the nucleon mass and $\mu=4.07\ \mathrm{MeV}^{\frac{1}{2}}$.

| n | $r_n$ | n | $r_n$ | n | $r_n$ |
|---|---|---|---|---|---|
| 1 | 0.603 | 4 | 0.988 | 7 | 0.999 |
| 2 | 0.906 | 5 | 0.995 | 8 | 1.000 |
| 3 | 0.968 | 6 | 0.998 | 9 | 1.000 |

Table I

In Table I we exhibit the rate of convergence for the ratios $r_n=\|U_{n+1}(s)\|/\|U_n(s)\|$ at the three-particle boundstate energy $s_o$ which was found to be $s_o=-25.2$ MeV. The coupling constant $\lambda_A$ was taken to be $\lambda_A=0.513\ \mathrm{MeV}^{\frac{1}{2}}$. This value gives rise to a two-particle boundstate at -2.225 MeV. From this table we see that the convergence rate is reasonable enough for practical purposes. Also it is amusing to note that the ratio $|r_{n+1}-r_n|/|r_n-r_{n-1}|$ is roughly independent of n, so that one already can make a sensible guess of the limiting value of $r_n$ with only a few terms known. This was found to be a general feature for most of the interactions studied.

Fig. 2 exhibits the dependence of the scattering length as a function of the coupling constant $\lambda_A$ for various diagonal Padé approximants. In Fig. 3 we show the Padé approximants as a

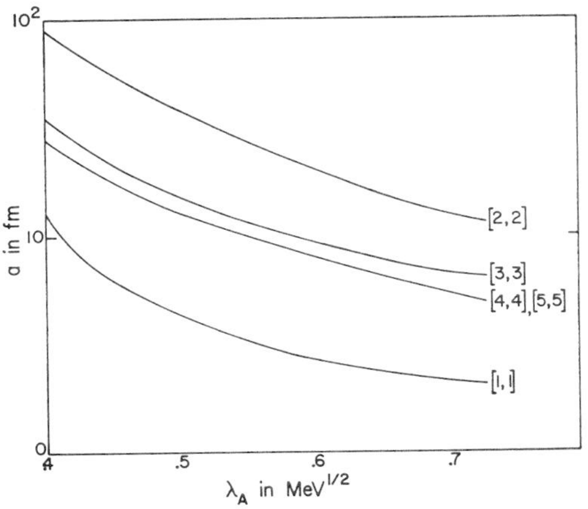


Fig.2.

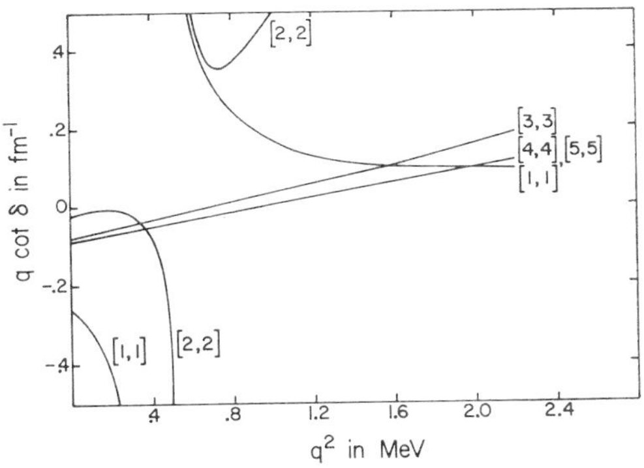


Fig.3.

function of the incoming free particle energy $q^2$ for the Yukawa potential with $\lambda_A$=0.513 $MeV^{\frac{1}{2}}$. In spite of the presence of possible poles in the approximants as a function of $q^2$ the convergence rate is satisfactory. Here about twelve terms in the multiple scattering series are needed to see the convergence. Finally, as an example for the convergence rate of the approximants in the inelastic region we show in Table II and III the Padé table

| D/N | 1 | 2 | 3 | 4 | 5 | 6 | 7 |
|---|---|---|---|---|---|---|---|
| 1 | -14.3 | -3.5 | -2.3 | -0.8 | -0.4 | -0.2 | -0.05 |
| 2 | 29.8 | 39.8 | -40.0 | 15.8 | 32.5 | 32.5 | 26.3 |
| 3 | 42.0 | 2.65 | 7.9 | 11.0 | 12.8 | 13.9 | 14.6 |
| 4 | -37.4 | 8.2 | 19.3 | 16.0 | 16.5 | 16.2 | 16.2 |
| 5 | -29.8 | 11.7 | 16.1 | 16.5 | 16.3 | 16.2 | 16.2 |
| 6 | -24.2 | 13.6 | 16.5 | 16.3 | 16.2 | 16.2 | 16.2 |
| 7 | -19.9 | 14.7 | 16.2 | 16.2 | 16.2 | 16.2 | 16.2 |

Table II

| D/N | 1 | 2 | 3 | 4 | 5 | 6 | 7 |
|---|---|---|---|---|---|---|---|
| 1 | 1.129 | 0.855 | 1.017 | 0.964 | 1.003 | 0.992 | 1.000 |
| 2 | 1.856 | 0.901 | 0.346 | 0.170 | 0.517 | 0.915 | 1.317 |
| 3 | 1.386 | 1.625 | 1.343 | 1.162 | 1.073 | 1.022 | 0.993 |
| 4 | 1.148 | 1.335 | 0.817 | 0.965 | 0.928 | 0.933 | 0.931 |
| 5 | 1.041 | 1.138 | 0.964 | 0.939 | 0.934 | 0.931 | 0.931 |
| 6 | 0.995 | 1.042 | 0.928 | 0.934 | 0.931 | 0.930 | 0.930 |
| 7 | 0.970 | 0.988 | 0.933 | 0.931 | 0.930 | 0.930 | 0.930 |

Table III

for the phaseshift $\delta$ (in degrees) and the inelasticity parameter $\eta$ respectively at $q^2$=10 MeV in the case of $\lambda_A$=0.513 MeV$^{\frac{1}{2}}$. From this we see that the final result is given by $\delta$=16.2$^{\circ}$ and $\eta$=0.930. There is little or no difference in the rate of convergence with respect to the elastic region. Of course at higher energies there should eventually be much better convergence, since the impulse approximation is expected to be valid at these energies.

## 6. Some concluding remarks

We have shown that the Pade technique can indeed be a practical tool for solving the three-body equations. Although we have demonstrated explicitly here its effectiveness only for a relatively simple potential, we have used the method quite successfully recently for studying more complicated interactions (Kloet et al., 1971). The restriction also to total angular momentum L=0 need of course not be made. For the higher partial waves it is found that the convergence rate is faster which is due to the presence of the centrifugal barrier. Also the method has been applied to the case of realistic two-body potentials (Reid, 1968, Bressel et al., 1969), which contain also tensor and spin-orbit components (Malfliet et al., 1970, 1971). Since the method is based only on the meromorphic character of the amplitude in the expansion parameter $\lambda$ its applicability is expected in general not to depend on the precise form of the interactions.

Up to now we have only considered the case of elastic scattering. Above the inelastic threshold it is in addition possible to have three free particles in the final state. The corresponding breakup amplitude can in principle also be determined using Padé approximants.

## References

Baker G. A., (1972). "The Existence and Convergence of Subsequences of Padé Approximants", preprint.

Bressel C. N., Kerman A. K. and Rouben B., (1969). Nucl. Phys. A124, 624.

Chisholm J. S. R., (1963). J. Math. Phys. 4, 12.

Faddeev L. D., (1961). Soviet Phys. JETP 12, 1014.

Faddeev L. D., (1965). "Mathematical Aspects of the Three-Body Problem in the Quantum Scattering Theory", Daniel Davey and Co. Inc., Hartford, Conn..

Kloet W. M. and Tjon J. A., (1971). Phys. Letters 37B, 460.

Kloet W. M. and Tjon J. A., (1972). "Elastic n-d scattering with local potentials", preprint.

Kowalski K. L., (1965). Phys. Rev. Letters, 15, 798.

Malfliet R. A. and Tjon J. A., (1969a). Nucl. Phys. A127, 161.

Malfliet R. A. and Tjon J. A., (1969b). Phys. Letters 29B, 293.

Malfliet R. A. and Tjon J. A., (1970). Ann. Phys. 61, 425.

Malfliet R. A. and Tjon J. A., (1971). Phys. Letters 35B, 487.

Noyes H. P., (1965). Phys. Rev. Letters 15, 538.

Nuttall J., (1970). J. Math. Anal. Appl. 31, 147.

McKee J. S. C. and Rolph P. M., (1970). In "Three-Body Problem in Nuclear and Particle Physics" (J. S. C. McKee and P. M. Rolph, eds.), North Holland Publ. Co., Amsterdam.

Reid R. V., (1968). Ann. Phys. 50, 411.

Tjon J. A., (1970). Phys. Rev. D1, 2109.

# PADE APPROXIMANTS IN POTENTIAL SCATTERING

C. R. Garibotti

(Istituto di Fisica dell'Università, Bari, Italy)

## 1. Introduction

Two body potential scattering is the usual testing ground for any approximation method intended to be useful in relativistic or in three body theories. We give here a partial account of the applications of Padé Approximants (P.A.) in potential theory.

First, we discuss the solution of the Lippmann Schwinger equation in a finite subspace and its relation with the Padé approximation. The results for the scattering amplitude follow from this relation. Then, we report the known results about unitarity of the P.A. and the convergence of the denominator of the [N,N] approximant to the Jost function. Part 3 contains a discussion of the approach used to calculate the binding energies with the P.A.. We describe the method used to resolve the difficulties found in the calculation of the poles of the on-shell scattering amplitudes. Furthermore we describe the limits of applicability of the perturbative expansion of the binding energies and the use of the P.A. as a summation method. We show the results obtained for the anharmonic oscillator and give the relation of the Brillouin Wigner perturbative method with series of Stieltjes.

Finally, we give a simple example of the use of P.A. to obtain the unphysical singularities of the S matrix in the k-plane.

## 2. The Wave Function

To study potential theory we can start either from the Schrödinger or from the Lippmann Schwinger integral equation. The last is more convenient because it contains implicitly the boundary conditions. The equations for the outgoing and ingoing wave functions

$$|\Psi^{(\pm)}(k)\rangle = |k\rangle + \lambda\, G_o^{(\pm)}(E)V|\,\Psi^{(\pm)}(k)\rangle \quad (1)$$

or for the standing wave function

$$|\Psi^{P}(k)\rangle = |k\rangle + \lambda\, G_o^{P}(E)V|\,\Psi^{P}(k)\rangle \quad (2)$$

$G_o^{(\pm)}(E) = (E-H_o \pm i\varepsilon)^{-1}$; $G_o^{P}(E) = P(E-H_o)^{-1}$; can be transformed into an equation with $L^2$ kernel. Sugar and Blankenbecler (1964) have attempted to introduce systematically transformations of this kind. Here we consider only the following reduced kernels

$$A^{(1)} = V^{\frac{1}{2}}\, G_o(E)\, V^{\frac{1}{2}}\, ,\quad A^{(2)} = G_o^{\frac{1}{2}}(E)\; V\; G_o^{\frac{1}{2}}(E)\ . \quad (3)$$

The kernel $A^{(1)}$ is $L^2$ in the whole physical sheet for almost all physically interesting potentials. Then, the Born expansion of the Eqs. 1 or 2 will have a finite radius of convergence in the $\lambda$ plane and the main use of P.A. will be the analytic continuation of the solution. The above equations are formally identical in the plane wave or in the angular momentum representations. Transforming accordingly $|\Psi^{(+)}\rangle$ and $|\phi\rangle$, the equations (1) can be written in a closed form

$$(1 - \lambda A)\; |\psi\rangle = |\phi\rangle\ , \quad (1 - \lambda A^{+})|\psi'\rangle = |\phi'\rangle\ . \quad (4)$$

The adjoint equation is introduced because physical kernels are usually non-hermitian. The solutions can be formally written in terms of the resolvent

$$|\psi\rangle = R(\lambda A)\,|\phi\rangle = (1-\lambda A)^{-1}\,|\phi\rangle\,, \quad |\psi'\rangle = R^{+}(\lambda A)\,|\phi'\rangle\,. \tag{5}$$

Starting from two vectors $|f\rangle$ and $|g\rangle$, we define an N-dimensional dual Hilbert subspace $(E_N, E_N^*)$ using the first N iterated vectors

$$|f_n\rangle = A^n\,|f\rangle \qquad |g_n\rangle = A^{+n}\,|g\rangle \quad (n=0,1,\ldots,N-1). \tag{6}$$

In each set the vectors are considered linearly independent and a biorthogonal base $\{|u_i\rangle, |u_i'\rangle\}$ can be obtained as linear combinations of the $f_n$ and $g_n$ by requiring

$$\langle u_i'|u_j\rangle = \delta_{ij}\,.$$

Then we obtain the projection operator onto the dual subspace

$$P_N(f,g) = \sum_{k=o}^{N-1} |u_k\rangle\langle u_k'|$$

and the projected kernel

$$A_N = P_N\,A\,P_N\,.$$

The corresponding resolvent results in (Garibotti, 1972)

$$R_N(\lambda A_N, f, g) = (1-\lambda A_N)^{-1} = D_N^{-1}(\lambda)\sum_{i=o}^{N-1} \lambda^i\,A_N^i\,S_{N-i-1}(\lambda) \tag{7}$$

with

$$D_N(\lambda) = \det{}_N(1-\lambda A_N) = \sum_{i=o}^{N-1} \lambda^i\,d_i$$

$$S_{N-\ell-1}(\lambda) = \sum_{i=o}^{N-\ell-1} d_i\,\lambda^i\,. \tag{8}$$

It provides solutions $|\psi\rangle_N$ and $|\psi'\rangle_N$ of the equations (1) in the dual subspace. These solutions are exact for the projected problem and can also be obtained by calculating the [N,N-1] approximant to the iterative expansion of $|\psi\rangle_N$ (Chisholm, 1963)

$$|\psi\rangle_N = |\phi\rangle + \lambda A_n |\phi\rangle + \lambda^2 A_N^2 |\phi\rangle + \ldots$$

that is

$$|\psi\rangle_N = R_N(\lambda A_N, f, g) |\phi\rangle = [N,N-1]_{|\psi\rangle_N} .$$

This approximation depends on the starting vectors f and g. In particular, the poles of $|\psi\rangle_N$, in the $\lambda$ variable are the zeros of $D_N(\lambda)$ and are independent of $\phi$ and $\phi'$. Furthermore, in $D_N(2)$ we can replace $A_N$ by A because

$$D_N(\lambda) = \det_N(1-\lambda A_N) = \det_N(P_N(1-\lambda A)P_N) = \det_N(1-\lambda A) \tag{9}$$

where we leave the subscript N to mean that the determinant must be calculated between the N vectors of the dual subspace. The difficulty is that the P.A. must be calculated with the finite kernel $A_N$ and it seems necessary to carry out the biorthogonalisation process and to obtain the projector. We could think of calculating the P.A. directly to the perturbative expansion of $|\psi\rangle$, that is $[N,N-1]_{|\psi\rangle}$, but the poles of these new approximants will not necessarily converge to the poles of the $\Psi$ (Chisholm, 1970). This difficulty of a straightforward application of the P.A. can be resolved by using the minimal iterations method. It consists essentially of making a particular choice of the starting vectors, i.e. $f=g=\phi$. Since the iterated vectors (6) are obtained by iterating with A and since all the vectors of the $|\Psi\rangle_N$ are in the subspace, we can forget about projection operators and write

$$|\psi\rangle_N = R_N(\lambda A, \phi, \phi) |\phi\rangle . \tag{10}$$

Now it can be shown that this $|\psi\rangle_N$ converges to $|\psi\rangle$ in the region where A is Hermitian (Nuttall, 1967), and, because the approach presented above can be analytically continued, the convergence can be proved outside this region (Garibotti and Villani, 1969a). The zeros of $\det_N(1-\lambda A)$ tend to the exact poles of $\psi$, this determinant being the denominator of the $[N,N-1]$ approximant to the series expansion of $\langle\phi|\psi\rangle$, as can be seen from the relation (Garibotti, 1972)

$$\langle g|R_N(\lambda A,f,g)|f\rangle = [N,N-1]_{\langle g|R(\lambda A)|f\rangle} . \tag{11}$$

Recalling that for Hermitian A, Masson (1970), using the spectral theorem, shows that

$$\langle f|R(\lambda A)|f\rangle = \int_{-\infty}^{\infty} \frac{d\phi(u)}{1-\lambda A} \tag{12}$$

$\phi(u)$ being a bounded monotonic non-decreasing function, we return to the physical problem.

## 2. The Scattering Amplitude

Considering the kernel $A^{(1)}$, the scattering amplitude can be written

$$\langle k'|T|k\rangle = \lambda\langle k'|V|\psi^{(+)}(k)\rangle = \lambda\langle k'|V^{\frac{1}{2}}\, R(\lambda A^{(1)})\, V^{\frac{1}{2}}|k\rangle . \tag{13}$$

For the on-shell forward scattering amplitude or for the partial-wave amplitude $|k\rangle \equiv |k'\rangle$, and assuming $V>0$ and $E<0$ the kernel $A^{(1)}$ is operator Hermitian and negative definite; then from Eq. (12)

$$\langle k|T|k\rangle = \lambda \int_{-\infty}^{0} \frac{d\phi(u)}{1-\lambda u} = \lambda \int_{0}^{\infty} \frac{d\phi(v)}{1+\lambda v} . \tag{14}$$

With the standing Green function for potentials of definite sign and E real, the kernel $A^{(1)}$ is Hermitian and the diagonal elements of the K matrix are

$$\langle k|K|k\rangle = \lambda\ \langle k|V|\psi^{P}(k)\rangle = \varepsilon\lambda \int_{-\infty}^{\infty} \frac{d\psi(u)}{1+\lambda u} \tag{15}$$

($\varepsilon$=sgn V). Meanwhile, by using the second kernel

$$\langle k'|T|k\rangle = \lambda \langle k'|V|k\rangle + \lambda^2 \langle k'|VG_0^{\frac{1}{2}} R(\lambda A^{(2)})G_0^{\frac{1}{2}} V|k\rangle . \tag{16}$$

For negative energy, $G^{\frac{1}{2}}$ is well defined, $A^{(2)}$ is $L^2$ and Hermitian and Eq. (12) gives

$$\langle k|T|k\rangle = \lambda \langle k|V|k\rangle + \lambda^2 \int_{-\infty}^{\infty} \frac{d\bar{\phi}(u)}{1+\lambda u} . \tag{17}$$

The functions $\phi$, $\psi$ and $\bar{\phi}$ are bounded and monotonic non-decreasing in the integration range. Then, the diagonal matrix elements of the K and T matrices are normal or extended series of Stieltjes (S.S.). When the reduced kernel of the scattering equations is bounded, the extended S.S. we have found can be reduced to normal S.S. by a linear fractional transformation (see the example in Baker, 1965). Since we are considering $L^2$ kernels, this is always true and the convergence properties of P.A. can be used.

From (15), the K matrix seems the more convenient quantity to deal with in the scattering problem. However, the analyticity of the scattering amplitude and of the eigenvalues in the cut physical energy sheet will allow us to continue our results to positive values of the energy.

To apply the convergence and bounding criteria of P.A. to nondiagonal matrix elements, we can use either the matrix method proposed by Nuttall (Baker, 1970) or reduce the problem to the calculation of diagonal matrix elements (Masson, 1970)

$$2\langle k'|T|k\rangle = (\langle k|+\langle k'|)T(|k\rangle +|k'\rangle) + (i-1) \langle k|T|k\rangle$$
$$- i(\langle k|+i\langle k'|)T(|k\rangle +i|k'\rangle) + (i-1) \langle k'|T|k'\rangle .$$

The P.A. to the non-diagonal elements can also be obtained from the subspace solution. If we get the dual base $\{|u_i\rangle , |u_i'\rangle\}$ using the vectors $|f\rangle = V^{\frac{1}{2}}|k\rangle$, $|g\rangle = V^{\frac{1}{2}}|k'\rangle$ the solution in the subspace gives the diagonal P.A. for the amplitude using (13) and (11)

$$\lambda \langle V^{\frac{1}{2}}k' | R_N(\lambda A^{(1)}, V^{\frac{1}{2}}k, V^{\frac{1}{2}}k') | V^{\frac{1}{2}}k \rangle = [N,N]_{\langle k'|T|k\rangle} . \quad (18)$$

The equivalence between P.A. and solutions in a finite subspace shows that the P.A. to the scattering amplitude could be considered as a particular choice in the separable potential formalism (Gillespie, 1967). Different people have discussed this relation (Tani, 1965; Chen, 1967; Garibotti and Villani, 1969c). An interesting example has been given by Nuttall (1967) where the bounding properties of the [N,N] approximants to the K matrix are obtained as a consequence of the bounds on the separable potential.

Now we can study the physical meaning of the denominator of the P.A.. Consider the partial wave representation, the [N,N] approximant is given by (18) with k=k'. As we have said before, the denominator of this approximant is $\det_N(1-\lambda A^{(1)})$ and because the kernel is completely continuous it converges to the Jost function (Garibotti and Villani, 1969b)

$$\lim_{N\to\infty} \det_N(1-\lambda A^{(1)}) = \det(1-\lambda V^{\frac{1}{2}}G_o V^{\frac{1}{2}}) = \det(1-\lambda G_o V) = f_\ell(E) \quad (19)$$

In any finite order, $\det_N(1-\lambda A^{(1)})$ will present the characteristic left-hand singularities of the perturbative terms but at the limit they must cancel out and only the right cut will remain. That could suggest that the P.A. violate unitarity to a finite order. However, from the rules of transformation of the P.A. under inversion and conjugation, we obtain (Bessis and Pusterla, 1962; Zinn-Justin, 1971) that

$$[N,M]^*_{S_\ell}(\lambda,k) = [M,N]^{-1}_{S_\ell}(\lambda,k)$$

where

$$S_\ell(\lambda,k) = 1 - \frac{2i}{k} \langle k|T(E)|k\rangle .$$

Then, the diagonal approximants to the S-matrix are unitary ($k^2=E$ positive real). For the amplitude, a more general unitarity relation (Caser et al., 1969) is true

$$\mathrm{Im}\ [N,M]\ \langle k|T|k\rangle = -\frac{2}{k}[N,M]^*\ \langle k|T|k\rangle \cdot [N,M]\ \langle k|T|k\rangle$$

when $N<M$. Then, even certain non-diagonal P.A. to the scattering amplitude are unitary. Because the P.A. to real quantities are real

$$\mathrm{Im}\ [N,M]\ \langle k|K|k\rangle = 0$$

for any N and M. Meanwhile, the P.A. to the total scattering amplitude are in general non-unitary.

## 3. Bound States

To obtain the energies of the bound states of $H=H_o+\lambda V$ we calculate the total or partial-wave scattering amplitude and then look for its poles at negative energies or we can use a direct perturbative formulation for these energies.

(i) Let us write the explicit integral equation for the scattering amplitude

$$T(k,k';E,\lambda) = V(k';k) + \lambda\int \frac{V(k,q)\ T(q,k';E,\lambda)}{E-q^2-i\varepsilon}\,\rho(q)\,dq \quad .$$

As before, k,k' and q must be intended as scalar or vector variables according to whether we deal with partial wave or total amplitude. By iterating this equation we get an expansion in $\lambda$ from which the P.A. are obtained. The N poles of the $[N,M]$ approximant, $\lambda^{(n)}=\lambda^{(n)}_{[N,M]}(k,k';E)$, $(1\leq n\leq N)$, are intended to provide approximate values for the N first bound states $E^{(n)}=E^{(n)}_{[N,M]}(k,k';\lambda)$ of the Hamiltonian H. As we see, the $E^{(n)}_{[N,M]}$ will depend on the initial and final momenta k and k'; such dependence must not appear at the exact energy. In particular, if we are using the total amplitude the poles will have an extra dependence in the momentum transfer $\Delta=|\vec{k}'-\vec{k}|$; they are usually called poloids and there is no rule to define a best choice for the value of $\Delta$. A variational approach, such as will be described later, could be possible, but its numerical realisation seems quite difficult. In this regard the minimal iterations method could be useful because its denominator has no spurious dependence in $\Delta$, see (9) and (10). The difficulties in the calculation of higher perturbative orders, the non-unitary of its P.A., the extra dependence in $\Delta$ of its poles and the poor results obtained with the on-shell values (Caser et al., 1969) made the total amplitude not a very convenient quantity for this kind of calculation.

The P.A. for the partial wave amplitude retains the spurious dependence in k and k'. When these values are chosen on-shell, i.e. $k=k'=E^{\frac{1}{2}}$, the denominator of the [N,N] approximant converges to the Jost function, but to any finite order the left-hand singularities of the perturbative terms in the E-plane provide divergent values of $\lambda^{(n)}_{[N,M]}$ for some values of E. For the exponential potential, that causes breakdown of the [M,1] and [M,2] approximants in the first and second left-hand poles respectively and so on (Basdevant and Lee, 1969). In general, that will happen in the threshold of the left cuts of the perturbative orders (Caser et al., 1969). Since the existence of left-hand singularities in the S-matrix are mainly due to the unboundedness of the initial and final states for complex values of the momenta, by choosing real values of k and k' that kind of difficulty should be avoided. A definite way to define these values has been proposed by Caneschi and Jengo (1969). They suggest using the relation between P.A. and variational principles (Nuttall, 1967; Garibotti, 1972) to consider the off-shell momenta k and k' as real parameters defined by stationary conditions. Alabiso et al. (1970, 1971a, 1971b) have made a systematic application of these ideas to calculate binding energies and phase shifts. If in some cases they found it necessary to introduce additional stability requirements, the stationary conditions giving an optimal value for $k=k'=\bar{k}$. Eliminating left-hand singularities in the denominator of the P.A., this method provides a very fast convergence of the $E^{(n)}_{[N,M]}(\lambda,\bar{k})$, to the correct value, even for very high values of $\lambda$. In all the considered cases the $[2,2]$ approaches the first S and P wave bound states. For other bound states a higher P.A. would be necessary.

(ii) So far, we have dealt with perturbative expansions of the integral equation, that is the Born series. Now we will consider the direct application of the perturbative methods to the discrete eigenvalues of the Schrödinger equation. Actually, there is a well established mathematical base to these approaches (Kato, 1966).

Consider the Hamiltonian $H(\lambda)=H(0)+\lambda V$, where $H(0)$ is an essentially self-adjoint operator and V is a regular perturbation in the sense that two non-negative constants a and b exist such that

$$\langle V\phi|V\phi\rangle \leq a \ \langle H(0)\phi|H(0)\phi\rangle + b \ \langle\phi|\phi\rangle \quad . \qquad (20)$$

Rellich (1953) has shown that if $H(0)$ has an m times degenerate eigenvalue $E_o$, there exist m eigenvalues $E_j(\lambda)$ of $H(\lambda)$ which are branches of algebroidal functions of $\lambda$, and the perturbation series converges for small enough values of $\lambda$. In practice the application of perturbation theory depends on the possibility of assigning to $H(\lambda)$ an invariant domain common with that of $H(0)$. This roughly means that the spectrum of $H(0)$ must not be destroyed when we turn on the interaction. Thus the validity of the perturbative expansion rests on the gentleness of the perturbation. When $H(\lambda)$ has an unambiguous essentially self-adjoint extension, the eigenvalue problem has a solution. Even when V is not a regular perturbation, but satisfies certain conditions, Kato has shown that the perturbative series has an asymptotic character. Then we can suppose that the perturbation terms carry the physical information, which can be found again by using an appropriate summation method. The P.A. has been applied with this aim. There are two kinds of perturbative approach, the explicit and the implicit. In the first kind we found the classical Rayleigh Schrodinger (R.S.) expansion and it is obtained by eliminating the eigenvalue and getting an equation for the eigenfunctions. The Brillouin Wigner (B.W.) is an example of implicit expansion; by eliminating the eigenfunction, an equation is obtained for the eigenvalue. We will describe a successful application of the P.A. in these two approaches.

Rayleigh Schrödinger expansion: Loeffel et al. (1969) have used the P.A. to sum the divergent R.S. expansion for the energy levels $E_n(\lambda)$ of the one dimensional anharmonic oscillator

$$\left[ - \frac{d^2}{dx^2} + \frac{x^2}{4} + \frac{\lambda}{4} x^4 \right] \phi(x) = E(\lambda) \phi(x) \quad .$$

The perturbation interaction $\lambda x^4$ dominates H(0) even for small values of $\lambda$. To look for the explicit reason for the divergence $E_n(\lambda)$ can be analytically continued to the complex $\lambda$-plane (Bender and Wu, 1969; Simon, 1970). It is found that $E_n(\lambda)$ has a third-order branch point at $\lambda$=0. $E_n(\lambda)$ has an analytical continuation in the first sheet, of the three-sheeted surface, cut along the negative real axis. To resolve this difficulty it could be enough to consider $\lambda^{1/3}$ as expansion parameter. However in the second and third sheet (around arg $\lambda=270^\circ$ and arg $\lambda=-270^\circ$) there exist infinite branch cuts with circular shape which have a limit point at the origin. Then, $\lambda$=0 is an essential singularity and there exist infinite Riemann sheets. This singularity is responsible for the perturbative divergence. For a fixed positive real value of $\lambda$ the real positive values of $E_n(\lambda)$ in the different sheets give an energy level and there is level crossing in these branch points. That means that by analytical continuation, knowing $E_o(\lambda)$, it is possible to know the whole spectrum of the anharmonic oscillator (Bender and Wu, 1969).

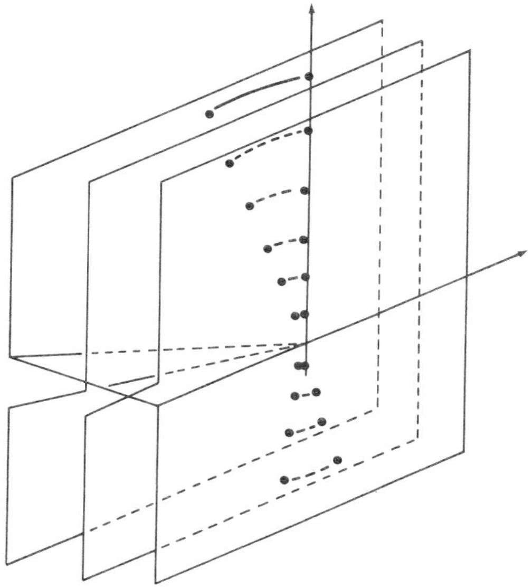

Fig.1. The Riemann sheet structure of $E_n(\lambda)$.

The analyticity of $E(\lambda)$ in the first cut sheet, the fact that Im $E(\lambda)$/Im $\lambda>0$ and certain bounds in the perturbative terms of $E(\lambda)$ around $\lambda=0$ show that $E(\lambda)$ is a S.S. and the $[N,N+j]$ approximants must converge for fixed j. Loeffel et al. (1969) have calculated up to order $[20,20]$. They find that the convergence of the P.A. is very fast for small values of $\lambda$ and slows down slightly for increasing $\lambda$. Most of the analytical study described above can be extended to the $\beta x^{2m}$ interactions and to the many-dimensional anharmonic oscillator (Simon, 1970) but a different summation method is necessary to get information from the perturbative expansion (Graffi et al. 1970).

The Brillouin Wigner Expansion: The Brillouin Wigner (B.W.) approach is another perturbative formalism to which we want to refer. Goscinski (Bendazzoli et al., 1970) has proved the relation of that method to the P.A., but we will give here a more succinct proof. We write

$$H\psi_n = (H_o + \lambda V)\,\psi_n = E_n\,\psi_n \qquad (21)$$

with

$$H_o\,\phi_n = E_n^{(o)}\,\phi_n$$

and considering only the ground state we have

$$\psi_o = \phi_o + \lambda\, PG_o(E_o)\, V\psi_o$$

where

$$P = 1 - |\phi_o\rangle\langle\phi_o| \,, \quad G_o(E) = (E_o - H_o)^{-1} \,.$$

Substituting in (21) and closing with $\phi_o$

$$E_o = E_o^{(o)} + \langle\phi_o|\lambda V(1 + \lambda\, PG_oV)^{-1}|\phi_o\rangle = E_o^{(o)} + \langle\phi_o|t(E_o)|\phi_o\rangle \,.$$

Developing in terms of $\lambda V$ we get the B.W. expansion. To prove the equivalence with P.A., we observe that the operator $t(E)$ satisfies the integral equation

$$t(E) = \lambda V - \lambda VPG_o(E)\, t(E) \,.$$

By similarity with the forward scattering amplitude, for negative values of E (in which we are interested), we can write

$$\langle\phi_o|t(E)|\phi_o\rangle = \lambda\,\langle\phi_o|V|\phi_o\rangle - \lambda^2 \int_{\infty}^{\infty} \frac{d\chi(u,E)}{1+\lambda u} \qquad (22)$$

for any potential which satisfies the usual regularity conditions. And

$$\langle\phi_o|t(E)|\phi_o\rangle = \lambda \int_{-\infty}^{\infty} \frac{d\phi(u,E)}{1+\lambda u}$$

for a repulsive potential. Where $\chi(u)$ and $\phi(u)$ are bounded non-decreasing functions. For a bounded potential, the extended S.S. in (22) can be reduced to a normal one. Then the P.A. constructed from the perturbative B.W. expansion will converge. The $[N,N]$ and $[N,N-1]$ approximants will provide upper and lower bounds for the bound state energies.

To perform calculations for the binding energies of the atoms in the isoelectronic sequence of He, Goscinsky has replaced the terms of the B.W. expansion, which are difficult to obtain, by those of the R.S. expansion. Calculations show that to obtain the rate of convergence of the P.A. with a R.S. expansion, it is necessary to calculate twice as many terms. Note that Kato (1951) has shown that for atomic problems the R.S. expansion has a non-zero radius of convergence. The P.A. work as a faster summation method.

## 4. Analytical Continuation in the k-Plane

We have shown the relation of the scattering amplitude and the bound state energy to series of Stieltjes when the coupling constant is used as an expansion parameter. We will study the analytical continuation of the scattering amplitude in the E-plane. Baker (1965) has used the Mandelstam representation to write the scattering amplitude as a series of Stieltjes in any of the three covariant variables. Here we will show a simple example of the behaviour of the P.A. in this case. Assume that the angular momentum is zero and use cut-off potentials

$$I(\alpha) = \int_0^\infty e^{\alpha r}|V(r)|\,dr < \infty \, .$$

In particular, we choose $V(r)=0$ for $r>a$. The Jost function is an entire function of k and the matrix $S(k)$ has no left-hand dynamical singularities in the E-plane. Because

$$S(k) \underset{|k|\to\infty}{\to} \frac{c}{k^2}\, e^{-2ika} \qquad (23)$$

in the upper half plane, there is an essential singularity at infinity. The singularities in the E-plane of $S_a(k) = S(k)\, e^{2ika}$ are the bound state poles and the kinematical cut. We can write a dispersion relation for

$$F(E) = \frac{S_a(E) - 1}{2ik} \quad ,$$

$$G(E) = \frac{F(E)-F(0)}{E} = \frac{1}{\pi}\int_0^{\infty} dE' \frac{\text{Im } F(E')}{(E'-E-i\varepsilon)E'} - \sum_n \frac{R_n/E_n}{E_n - E} = \int_{-\infty}^{\infty} \frac{d\psi(u)}{1-uE} \tag{24}$$

with

$$\text{Im } F(E) = k|F(E)|^2 > 0 \quad \text{and} \quad R_n^{-1} = -\int_0^{\infty} f^2(ik_n,r)\, dr < 0 \ .$$

$f(ik_n,r)$ is a real function, since $E_n=-k_n^2$ . $\psi(u)$ is a bounded non-decreasing function and $G(E)$ is an extended series of Stieltjes. The $[N,N+j]$ approximants with $j=\pm1,\pm3,\ldots$ will provide a convergent analytic continuation of $G(E)$ to the complex E-plane. That is not enough to obtain an extension onto the unphysical sheet, i.e. the lower half k-plane. Since $S(k)$ is a meromorphic function of k, we can use a theorem of Nuttall (1970): "The P.A. to a meromorphic function converge in measure to that function within a closed bounded region". Then we could use a low energy expansion of $S(k)$ to obtain with the P.A. an analytical continuation in the k-plane and look for complex poles. As an example, we have carried out this calculation for the square well potential. In the figures we show some of our results and the exact trajectories (Nussenzveig, 1959).

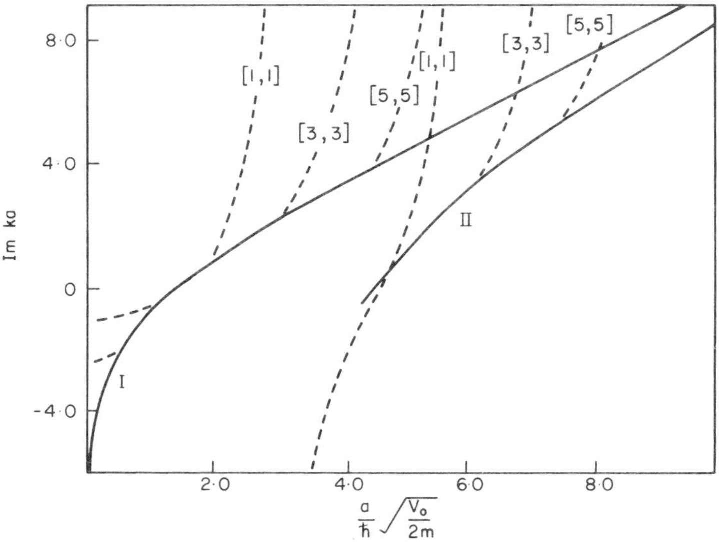


Fig.2. I, II and the dotted lines are pole trajectories for a square well potential s-wave amplitude and its P.As.

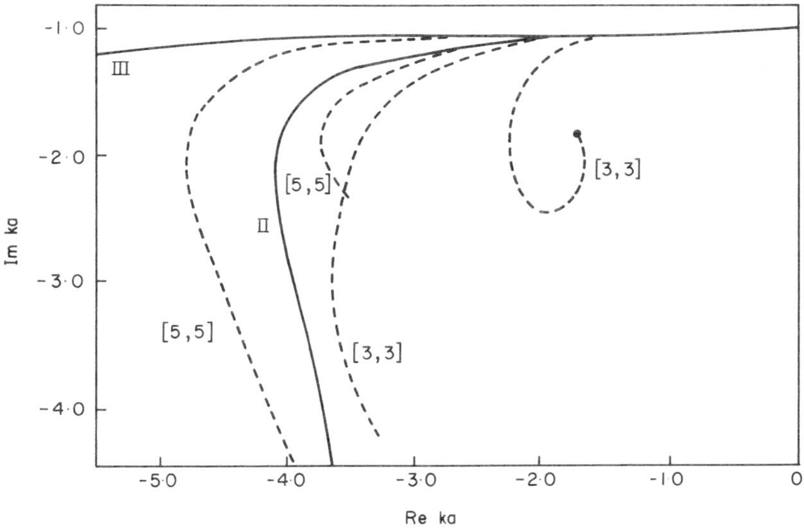


Fig.3. II, III and the dashed lines are the second and third pole trajectories for a square well potential s-wave amplitude and its P.As. in the complex k-plane.

References

Alabiso C., Butera P. and Prosperi G. M., (1970). Nuovo Cimento Lettere, 3, 831.

Alabiso C., Butera P. and Prosperi G. M., (1971a). Nucl. Phys., 31B, 141.

Alabiso C., Butera P. and Prosperi G. M., (1971b). Milano preprint.

Baker G. A., (1965). Adv. Theor. Phys., 1, 1.

Baker G. A., (1970). In "The Padé Approximant in Theoretical Physics" (G. A. Baker and J. L. Gammel, eds.), p.2, Academic Press, N.Y.

Basdevant J. L. and Lee B. W., (1969). Nucl. Phys., 13B, 182.

Bendazzoli G. L., Goscinski O. and Orlandi G., (1970). Phys. Rev., 2A, 2.

Bender C. and Wu T. T., (1969). Phys. Rev., 184, 1231.

Bessis D. and Pusterla M., (1968). Nuovo Cimento, 54A, 243.

Caneschi L. and Jengo R., (1969). Nuovo Cimento, 60A, 25.

Caser S., Piquet C. and Vermeulen J. L., (1969). Nucl. Phys., 14B, 119.

Chen A., (1967). Nuovo Cimento, 52A, 474.

Chisholm J. S. R., (1963). J. Math. Phys., 4, 1506.

Chisholm J. S. R., (1970). In "The Padé Approximant in Theoretical Physics" (G. A. Baker and J. L. Gammel, eds.), p.171, Academic Press, N.Y.

Garibotti C. R., (1972). Ann. Phys. (NY), 71, 446.

Garibotti C. R. and Villani M., (1969a). Nuovo Cimento, 59A, 107.

Garibotti C. R. and Villani M., (1969b). Nuovo Cimento, 61A, 747.

Garibotti C. R. and Villani M., (1969c). Nuovo Cimento, 63A, 1267.

Gillespie J., (1967). Phys. Rev., 160, 1432.

Graffi S., Grecchi V. and Simon B., (1970). Phys. Letters, 32B, 631.

Kato T., (1951). Trans. Am. Math. Soc., 70, 195.
Kato T., (1966). "Perturbation Theory for Linear Operators". Springer-Verlag (Berlin).
Loeffel J. J., Martin A., Simon B. and Wightman A., (1969). Phys. Letters, 30B, 656.
Masson D., (1967). J. Math. Phys., 8, 2308.
Masson D., (1970). In "The Padé Approximant in Theoretical Physics" (G. A. Baker and J. L. Gammel, eds.), p.197, Academic Press, N.Y.
Nussenzveig H. M., (1959). Nucl. Phys., 11, 499.
Nuttall J., (1967). Phys. Rev., 157, 1312.
Nuttall J., (1970). J. Math. Anal. Appl. 31, 147.
Rellich R., (1953). "Perturbation Theory of Eigenvalue problems". Lecture notes, New York University.
Simon B., (1970). Ann. Phys., 58, 76.
Sugar R. and Blankenbecler R., (1964). Phys. Rev., 136B, 472.
Tani S., (1965). Phys. Rev., 139B, 1011.
Zinn-Justin J., (1971). Phys. Reports, 1C, 55.

# PADE APPROXIMANTS AND THE LIPPMANN SCHWINGER EQUATION

P.R. Graves-Morris and J.F. Rennison

(Mathematical Institute, University of Kent, England)

## 1. Introduction

We wish to solve the partial wave Lippmann Schwinger equation for the K-matrix (PWLSK). This is

$$K(p,k') = V(p,k') - \frac{2}{\pi} \int_0^\infty V(p,q) \frac{q^2 dq}{k^2-q^2} K(q,k')$$

For a fixed positive value of k', this becomes

$$f(p) = \phi(p) + \frac{2}{\pi} \int_0^\infty V(p,q) \frac{q^2 dq}{q^2-k^2} f(q)$$

The normalisation is chosen so that, for $k' = p = k$, $f(k) = k^{-1} \tan \delta$; PWLSK is thus a real, singular integral equation, and its solution gives the partial wave phase shift $\delta$. We write the equation and its series solution formally as

$$K = V + VPGK$$
$$= V + VPGV + VPGVPGV \ldots\ldots$$

Let $V = gU$, where g is the coupling strength. Then $K = gU + g^2\, UPGU + g^3\, UPGUPGU \ldots\ldots$ and we can form Padé Approximants (P.As.) to this as a power series in g. The obvious questions are whether a solution of PWLSK exists and whether the P.As. converge to the solution. We will show that $(1 - gUPG)^{-1}$ exists and is meromorphic in g. The theorems of Beardon (1968), Nuttall (1970) and Pommerenke (1972) now determine what is known about the convergence of the P.As. to the K-matrix.

Various techniques with various conditions can be used to establish meromorphy, and a method which will generalise is to be preferred. One has in mind the Blankenbecler Sugar equation (1966) and the Bethe Salpeter equation. One wishes to be able to use local two sign potentials and exchange potentials. The methods which have been used so far are

(i) Hilbert space methods :-

$$K = V^{\frac{1}{2}}(1 - V^{\frac{1}{2}}PGV^{\frac{1}{2}})^{-1} V^{\frac{1}{2}}$$

When $V(r)$ is a short range potential, $\Psi(\underline{r}) = V^{\frac{1}{2}}(r)\psi(\underline{r})$ is an $L^2$ function and then $V^{\frac{1}{2}}PGV^{\frac{1}{2}}$ can be shown to be compact on the $L^2$ space, and the resolvent $(1 - V^{\frac{1}{2}}PGV^{\frac{1}{2}})^{-1}$ exists on the space and is meromorphic in g.

(ii) Subtraction and Symmetrisation methods :-
These are tricks which produce Fredholm equations, due to Taylor (1963), Kowalski (1965), Noyes (1965), Graves-Morris (1971) and Osborne (1971).

(iii) Poincaré's theorem applied to the Schrödinger equation shows that the Jost solutions and Jost functions for non-singular potentials are analytic in g, hence the S-matrix is meromorphic.

(iv) Banach space methods :-
Lovelace (1964) used the space $C_1$ of bounded, differentiable functions with bounded continuous derivatives on $[0,\infty)$, and considered for Im E > 0,

$$\hat{K}_\ell(E) = \frac{1}{\pi} \int_0^\infty dE' \frac{\hat{\Delta}_\ell(E')}{E'-E}$$

where $< p \mid \hat{\Delta}_\ell \mid q > = 2\pi^2 V_\ell(p,\sqrt{E})\sqrt{E} \;\; \delta(q-\sqrt{E})$ .

Then $\hat{\Delta}_\ell(E)$ is a rank one operator on the space $C_1$, and the integral can be approximated by a Riemann sum, which is a finite rank operator. Modification of a theorem of Muskhelishvili shows that, provided $\hat{\Delta}_\ell(E')$ has a derivative with respect to E' which is bounded in norm in the neighbourhood of any point on the positive real axis, $\hat{K}_\ell(E)$ converges in norm as Im E tends to zero to the kernel of PWLSK. Hence $\hat{K}_\ell(E)$ for E real and positive can also be approximated in norm by operators of finite rank.

## 2. A simpler Banach space method

We wish to avoid the use of operator valued integrals and to combine the techniques of the subtraction method with the Banach space approach. Define

$$(Tf)(p) = P\int_0^\infty V(p,q)\,\frac{q^2dq}{q^2-k^2}\,f(q)\,dq$$

$$= \int_0^\Lambda \frac{V(p,q)-V(p,k)}{q-k}\,\frac{q^2dq}{q+k}\,f(q)$$

$$+ V(p,k)\int_0^\Lambda \frac{q^2}{q^2-k^2}\,f(q)\,dq + \int_\Lambda^\infty V(p,q)\,\frac{q^2dq}{q^2-k^2}\,f(q)$$

$$= (T_1f + T_2f + T_3f)\,(p)\ .$$

We show that $T_1$, $T_2$ and $T_3$ are bounded operators which map $C_1$ functions into $C_1$ functions; $T_2$ is a rank one operator; $\Lambda$ is chosen so that $T_3$ is an operator of arbitrarily small norm; $T_1$ may be shown to be the sum of a finite rank operator and an operator of arbitrarily small norm. To do this we need conditions on the potential, similar to those used by Lovelace, (it is true (Graves-Morris and Rennison, 1972) that the Yukawa potential satisfies these) :-

1. $|V(p,q)| \leqslant \Phi(q)$ and $\left|\frac{\partial V}{\partial p}(p,q)\right| \leqslant \Phi(q)$

   for all $p,q \geqslant 0$ and some $\Phi(q)$ such that $\int_0^\infty \Phi(q)\,dq < \infty$

AND 2. $V(p,q)$ has continuous second partial derivatives for $p \geqslant 0$, $q \geqslant 0$. Also, for any $\Lambda > 0$, $\frac{\partial V}{\partial q} \to 0$, $\frac{\partial^2 V}{\partial p\,\partial q} \to 0$ as $p \to \infty$ uniformly for $o \leqslant q \leqslant \Lambda$.

Since $T = g(N + Q)$, where $Q$ is a finite rank operator and $N$ has arbitrarily small norm, $T$ is a compact operator, and we solve

$$f = \phi + 2\pi^{-1}\, Tf = \phi + 2g\pi^{-1}\,(N+Q)f$$
$$= (1-2g\pi^{-1}N)^{-1}\,\phi + 2\pi^{-1}g\,(1-2\pi^{-1}gN)^{-1}Qf.$$

Either by this explicit method, or because T is compact, f is shown to be meromorphic in g, and the theorem of Pommerenke shows that the P.As. to f converge in capacity.

## References

Beardon A. F., (1968), J. Math. Anal. Applic. 31, 147.
Blankenbecler R. and Sugar R., (1966), Phys. Rev. 142, 1051.
Graves-Morris P. R., (1971), Nuovo Cim. 4A, 91.
Graves-Morris P. R. and Rennison J. F., (1972) Kent preprint.
Kowalski K. L., (1965), Phys. Rev. Letters 15, 798.
Lovelace C., (1964), Phys. Rev. 135B, 1225.
Noyes H. P., (1965), Phys. Rev. Letters 15, 538.
Nuttall J., (1970), J. Math. Anal. Applic. 31, 147.
Osborne T. W., (1971), Phys. Rev. D3, 395.
Pommerenke Ch., (1972), J. Math. Anal. Applic., to be published.
Taylor J. G., (1963), Suppl. al Nuovo Cim. Vol.1, p.1002.

# PADÉ APPROXIMANTS IN QUANTUM FIELD THEORY

D. Bessis

(University of Western Ontario, Canada

and

Centre d'Etudes Nucleaires de Saclay, France)

## 1. Introduction

### 1-a. The problem

The equations for Field Theory are non-linear partial differential equations for operator valued functions: they describe systems with an infinite number of degrees of freedom. Up to now, no solution is known. Even an existence theorem is missing. Due to the extreme difficulty of the problem, only a formal solution is available. This solution presents itself as a formal series in the coupling strength whose $n^{th}$ Taylor coefficients is given by an n-uple integral. A first difficulty arises from the fact, that once properly redefined, the Taylor series appears to be of zero radius of convergence. It is therefore necessary to apply an accelerator of convergence to it: this will be the reason to introduce Padé Approximations. However, it will appear that the introduction of such approximations is linked with even deeper reasons, such as the analytic properties of the solution in the coupling strength on one hand and on the other hand the variational properties of these approximations.

## 1-b. The example of Quantum Electrodynamics

In the theory of photons and electrons, the usual perturbation approach (the formal renormalized Taylor Expansion to the scattering matrix) has given the most remarkable quantitative results. However, such results are connected with the following features:

- The photon field has a classical limit, therefore one can reconstruct the interaction Lagrangian through the correspondence principle.
- It is possible to redefine the Taylor coefficients of the S-matrix expansion, through the use of essentially only two physical parameters: the photon and electron mass (mass renormalization) and the strength (coupling constant renormalization).
- The formal S-matrix renormalized expansion, although very likely defining a series of zero radius of convergence, behaves like an asymptotic series: the first successive coefficients decrease exponentially.
- The problem of bound states is solved by the use of the gauge invariance group which allows one to choose the radiation gauge: here the Coulomb part of the field, which is the strongest part and for which an exact solution is known, is separated out explicitly. The remaining part of the interactions is weak and treated perturbatively.

The results reach a precision achieved only in Astronomy.

Due to those exceptional circumstances, the difficulty of

the perturbative approach does not fully appear.

## 1-c. The case of the strong interaction physics

In the case of the interaction between pions and nucleons, one has first of all to choose the type of interaction Lagrangian. This is made difficult because there is no classical limit to such interaction, and we cannot apply the principle of correspondence. The symmetry laws: Lorentz invariance, Isospin invariance, Chiral invariance, Current algebra requirements, etc. will of course reduce considerably the possible forms. To reduce even more the choice, one could ask for renormalizability of the Lagrangian. However this concept of renormalizability is time-dependent: a Lagrangian non-renormalizable nowadays can become renormalizable later on. In fact a Lagrangian such as the non-linear-$\sigma$-model which is still non-renormalizable, in the usual sense, may appear to be very good to describe the physics of pions and nucleons up to 1.5 Gev.

Therefore the problem for the search of the correct Lagrangian appears to be very important, and will necessitate the trial of many possible forms; some of them will be discussed in Section 4.

Besides those difficulties, one finds that most often the second perturbative approximation is much larger than the first and therefore the series has no more its asymptotic nature. This is connected with a simple physical fact: there is very often a resonance or a bound state in the corresponding channel, that is, a pole in the S matrix, which will make the series lose its asymptotic nature.

Finally, in the case of the strong interaction between pions and nucleons, there is no more gauge group and also no exact solutions to any approximate scheme are known.

### 1-d. The diverging nature of the solution and the problem of its analytical continuation

On the practical point of view of computing physical quantities there is a great difference between Quantum Electrodynamics and strong interaction pion-nucleon dynamics. On the fundamental point of view there is no difference: in both cases, one is faced with the problem of summing up a divergent series. It has been proved in many models, such as the Peres model (Peres, 1963), the $\phi^3$ and $\phi^4$ (Hurst, 1952) interactions, that the scattering matrix has in the complex plane of the strength parameter, a singularity at zero. To start with, we shall suppose some analyticity in the coupling constant, and look for a way to reconstruct the solution from the knowledge of its derivative in a singular point. It is clear that we must postulate uniqueness of the procedure, because one can always add a function with all derivatives equal to zero at the origin. The approximations involving rational fractions whose poles and zeros by clustering can rebuild cuts and by coaslescing rebuild essential isolated singularities, seem to be very suitable. Furthermore in such approximations, poles have become harmless singularities.

Among such rational fractions the Padé Approximants built from the knowledge of only a finite number of Taylor series coefficient, appear, due to their special covariance properties with respect to homographical transformations giving rise to exactly unitary solutions (Gammel and MacDonald, 1966) and to their variational properties (Nuttall, 1970; Bessis, 1972; Caneschi and Jengo, 1969; Bessis and Talman, 1972; Alabiso et al., 1971-1972) a better choice.

## 2. The Construction of the S-matrix expansion

### 2-a. The unitarity of the S matrix

We write, introducing the T matrix

$$S(g) = I + 2i\ T(g) \tag{2-a-1}$$

where $g$ is the expansion parameter (coupling constant).

Unitarity reads

$$S(g)\ S^{+}(g^{*}) = S^{+}(g^{*})\ S(g) = I \tag{2-a-2}$$

or, using the T matrix

$$\frac{T(g) - T^{+}(g^{*})}{2i} = T(g)T^{+}(g^{*}) = T^{+}(g^{*})T(g) \tag{2-a-3}$$

Expanding $T(g)$ in powers of $g$ we get

$$T(g) = \sum_{n=0}^{\infty} T_n g^n \tag{2-a-4}$$

and identifying in both members of (2-a-3) the power of $g^n$:

$$\frac{T_n - T_n^{+}}{2i} = T_1 T_{n-1}^{+} + T_2 T_{n-2}^{+} + \dots + T_{n-1} T_1^{+} \tag{2-a-5}$$

(2-a-5) tells us that the anti-hermitean part of $T_n$ is fixed by the knowledge of $T_1, T_2 \ldots T_{n-1}$. It remains to fix its Hermitean part. Furthermore for $n = 1$ we get:

$$T_1 = T_1^{+} \qquad (2\text{-a-}6)$$

That is, $T_1$ is Hermitean.

2-b. <u>The Hermitean part of $T_n$ and the renormalization</u>

We shall consider the problem of elastic two-body scattering of a scalar field for simplicity. The matrix element of the T operator between the initial and final state, can be expressed, making use of Lorentz invariance:

$$\langle p_1', p_2' | T | p_1, p_2 \rangle = \mathcal{T}(s,t,u,p_1^2,p_2^2,p_3^2,p_4^2) \qquad (2\text{-b-}1)$$

where $s = (p_1 + p_2)^2$; $t = (p_1 - p_1')^2$; $u = (p_1 - p_2')^2$ are the usual invariants connected by the relation

$$s + t + u = \sum_1^4 p_i^2 \qquad (2\text{-b-}2)$$

We shall suppose that $\mathcal{T}(s,t,u)$ is analytic in the tensor product of the s,t and u plane, cut from $4\mu^2$ to infinity (where $\mu$ is the mass of the particle in the field), but restricted by condition (2-b-2).

Such hypothesis (Mandelstam) is too restrictive, and the method we expose here can be generalized to include far less analyticity. For the simplicity and clarity of the exposure, we shall suppose to be in this simple case. This case includes in particular all calculations with usual Lagrangians, when stopped at low perturbative order. Due to the extreme difficulty to compute high order perturbation this includes the practical case.

Then the discontinuity of $\mathcal{T}_n(s,t,u,p_1^2,p_2^2,p_3^2,p_4^2)$ on the s-plane cut is given by the anti-hermitean element of matrix of $T_n$. Therefore, starting from $T_1 = T_1^+$ (2-a-6) we see that $\mathcal{T}_1(s,t,u,p_i^2)$ is necessarily a real meromorphic function. The simplest choice is

$$\mathcal{T}_1(s,t,u,p_i^2) = \sum_{\substack{\text{perm} \\ s,t,u}} \frac{g_i}{s-\mu_i^2} + \text{polynomial } (s,t,u) \qquad \text{(2-b-3)}$$

The physical interpretation of the pole $\frac{g_i}{s-\mu_i^2}$ is very simple: it corresponds to an elementary particle of mass $\mu_i$, which couples with strength $g_i$ to the particles of the field. The polynomial also has a simple interpretation: it corresponds to a direct interaction among four particles of the field, involving derivative coupling.

From the relation:

$$\frac{T_2 - T_2^+}{2i} = T_1^2 \qquad \text{(2-b-4)}$$

we can compute the discontinuity of $\mathcal{T}_2$ in the s plane, for instance:

$$\text{Im}_s \ \mathcal{T}_2(s,t,u,p_i^2) = \int \mathcal{T}_1 * \mathcal{T}_1 \, dn_2 \qquad \text{(2-b-5)}$$

where $dn_2$ is the element of phase space corresponding to two particles; the integral is of course convergent, because it is taken over a compact. For a scalar field, crossing symmetry tells us that $\mathcal{T}(s,t,u,p_i^2)$ is invariant under any permutation of $s,t,u$. This allows us to deduce from (2-b-5) $\mathrm{Im}_t\ \mathcal{T}_2(s,t,u,p_i^2)$ and $\mathrm{Im}_u\ \mathcal{T}_2(s,t,u,p_i^2)$.

Knowing the three discontinuities of $\mathcal{T}_2(s,t,u,p_i^2)$ on its cuts, there is no difficulty to reconstruct $\mathcal{T}_2$. However, we can always add to $\mathcal{T}_2$ an arbitrary entire function, we shall always construct a "minimal" solution which seems to be the only one which makes sense physically. If the polynomial which is in $\mathcal{T}_1$ is not a constant, $\mathrm{Im}_s\ \mathcal{T}_2$ will behave at least like $s^2$ at infinity, and the theory will involve automatically an infinite number of parameters, because at next order $\mathrm{Im}_s\ \mathcal{T}_3$ behaves like $s^3$ at least, and so forth. When the polynomial is reduced to a constant, one can show that $\mathrm{Im}_s\ \mathcal{T}_n(s,t,u,p_i^2) \propto (\log s)^{n-2}$ for $s \to \infty$ and therefore only one subtraction is needed at each order to reconstruct by dispersion relations a "minimal" solution. This last case corresponds to renormalizable Lagrangians. However, even in the case where the polynomial in $\mathcal{T}_1$ is not a constant, which seems to correspond to the real physical case, because one limits himself to a finite order, only a finite number of parameters,enter in the solution. It is necessary that those parameters be correlated, and not adjusted to fit the experimental results. This is so, because it is only if the successive iterations are strongly correlated that we can expect

P. A. constructed out of them to converge. Therefore it is necessary to have physical prescriptions to generate the subtractions, (Renormalization).

## 3. Properties of the S-matrix Padé solution

### 3-a. The Unitarity and Analyticity Properties

From the knowledge of the formal expansion to the T matrix:

$$T = T_1 + T_2 + \dots T_n + \dots \tag{3-a-1}$$

We can construct in the Hilbert space corresponding to the two particles the first Padé Approximation:

$$[1/1]_T = T_1[T_1 - T_2]^{-1} T_1 \quad . \tag{3-a-2}$$

This operator, as well as the $[N/M]_T$ with $M \geq N$ are automatically unitary, as P. A. to unitary operators. Nevertheless, it is very interesting to diagonalize them partially by reducing all the $T_n$ under the rotation group. Due to the fact that the P. A. of a direct sum of operators is the direct sum of the P. A. on each term, we lose no information at all, and the operators are much simpler.

We shall consider the partially diagonalized $T$:

$$T^J = T_1{}^J + T_2{}^J + \dots T_n{}^J + \dots \tag{3-a-3}$$

in the case where $T_1{}^J = 0$ (which accidentally happens with the $\lambda\phi^4$ Lagrangian), one has to consider the next appproximations.

$$
\begin{aligned}
[1/1]^J_T &= T^J_1 [T^J_1 - T^J_2]^{-1} T^J_1 \quad \text{unitary} \\
[1/2]^J_T &= T^J_2 [T^J_2 - T^J_3 - 2i(T^J_2)^2]^{-1} T_2 \quad \text{unitary} \\
[2/1]^J_T &= T^J_2 [T^J_2 - T^J_3]^{-1} T^J_2 \\
[2/2]^J_T &= T^J_2 [T^J_2 - T^J_3 + T^J_3 (T^J_2)^{-1} T^J_3 - T^J_4]^{-1} T_2{}^J \quad \text{unitary.}
\end{aligned}
\tag{3-a-4}
$$

If we consider a two-body scattering we can set, by going to the C. M. S.:

$$
\begin{aligned}
p_1 &= (\frac{\sqrt{s}}{2} + w,\ p\hat{p}) \\
p_2 &= (\frac{\sqrt{s}}{2} - w,\ - p\hat{p}) \\
p_3 &= (\frac{\sqrt{s}}{2} + w',\ p'\hat{p}') \\
p_4 &= (\frac{\sqrt{s}}{2} - w',\ - p'\hat{p}')
\end{aligned}
\tag{3-a-5}
$$

Therefore $T^J_n$ are operators whose matrix elements are

$$
\langle w' p' | T^J_n(s) | wp \rangle \tag{3-a-6}
$$

The on-shell conditions are obtained setting:

$$
\begin{aligned}
p_1{}^2 &= p_3{}^2 = m^2 \\
p_2{}^2 &= p_4{}^2 = \mu^2
\end{aligned}
\tag{3-a-7}
$$

that is:

$$w = w' = \frac{m^2 - \mu^2}{2\sqrt{s}}$$

$$p = p' = \frac{\sqrt{[s-(m+\mu^2)][s-(m-\mu^2)]}}{\sqrt{4s}} \quad . \tag{3-a-8}$$

The analytic properties in s of the $T_n^J(s)$ operators, will be automatically preserved by the P. A. because the matrix elements of the P. A. are rational fractions built up through multi-linear expressions of the matrix element of the $T_n^J(s)$. (This is true on-shell or off-shell).

It is in general a formidable task to compute the inverse of $(T_1^J - T_2^J)$, for instance, to construct the [1/1] P. A. In particular, before any inversion, it would be good to know if such inverse exists. There are two very important cases in which it is the case: the Schroedinger equation, and its generalization: the ladder approximation to the Bethe-Salpeter equation. The reason for this is simple: the [1/1] P. A. is, in those cases, the exact solution. More generally the existence of the inverse is linked with the existence of a Quasi-potential, because it can be shown that the [1/1] P. A. is the solution for the Quasi-potential. The advantage of the [1/1] P. A. over the Quasi-potential is practical: it short-cuts its construction. In the case of the $\lambda\phi^3$ model, it would be nice to study the existence from the knowledge of the matrix elements (3-a-6). Such study remains to be done.

3-b. The mass spectra

Let us study the simplest case: the [1/1] P. A. reads, setting down the coupling explicitly:

$$[1/1]^J_T(s) = \lambda T^J_1(s)[T^J_1(s) - \lambda T^J_2(s)]^{-1} \, T^J_1(s) \tag{3-b-1}$$

There exist values of $\lambda$ for which $T^J_1(s) - T^J_2(s)$ has no inverse, those values $\lambda = \lambda^J(s)$ define the bound states of angular momentum $J$ of the problem, if $s$ is real and between 0 and the first threshold $(m+\mu)^2$ or resonances if $s$ is in a second sheet, $\mathrm{Re}\, s > (m+\mu)^2$ and $\mathrm{Im}\, s < 0$ and small. More generally we can define extension to $J$ continuous or complex of (3-b-1) by using Froissart-Gribov integrals to define the matrix elements (3-a-6) and obtain the Regge trajectories. Therefore the mass spectra, or generalized mass spectra come out very naturally through the denominators of the P. A.

3-c. The Variational principles of Kohn and Lippman-Schwinger

It can be shown that the diagonal Padé Approximant to the evolution operator are solution of the Lippman-Schwinger (1950) Variational principle, while the [N+1/N] are solution of the Kohn Variational principle (Kohn, 1947). The Ansatz to use, associated with those principles is the Cini-Fubini Ansatz (Cini and Fubini, 1956). More precisely if

$$U(t_2;t_1) = \sum_0^\infty U_n(t_2;t_1) \tag{3-c-1}$$

where $\sum_0^\infty U_n(t_2;t_1)$ is the Liouville iterated series of the equation:

$$U(t;t_1) = I - i\int_{t_1}^{t} H(t')\,U(t';t_1)\,dt' \qquad (3\text{-c-}2)$$

One can construct the [N/N] P. A. from (3-c-1).

On the other hand, $U(t_2;t_1)$ is the stationary value of the functional

$$F(V^+;V;t_2;t_1) = I - i\int_{t_1}^{t_2} dt[V^+(t;t_2)H(t) + H(t)\,V(t;t_1)]$$

$$+ i\int_{t_1}^{t_2} V^+(t;t_2)H(t)V(t;t_1)\,dt \qquad (3\text{-c-}3)$$

$$+ i^2\int_{t_1}^{t_2} dt\int_{t_1}^{t_2} dt'\; V^+(t;t_2)H(t)\theta(t-t')V(t';t_1)$$

Setting:

$$V^{(N)}(t;t_1) = \Lambda_1 + U_1(t;t_1)\,\Lambda_2 + \dots + U_{n-1}(t;t_1)\,\Lambda_n \qquad (3\text{-c-}4)$$

$$V^{+(N)}(t;t_2) = M_1 + M_2U_1^{\;+}(t;t_2) + \dots + M_nU_{n-1}^{+}(t;t_2) \qquad (3\text{-c-}5)$$

where the $\Lambda_i$ and the $M_j$ are time independent operators. If one looks for the stationary value of $F$ with such an Ansatz, one finds precisely the [N/N] P. A. to $U(t_2;t_1)$.

Those properties show that we can expect rapid convergence from P. A. In particular, while our P. A. do not satisfy exactly crossing symmetry, being variational solutions, they will violate crossing only variationally, while being exactly unitary. (It can be shown that exactly unitary approximations involving only a finite number of coupled channels must violate crossing, otherwise they would be identically zero).

3-d. Practical convergence

As we have seen, the [1/1] P. A.

- is exactly unitary
- has the correct analytic properties in energy
- is the exact solution of any Schroedinger-like equations
- is the simplest solution of the Lippmann-Schwinger variational principle
- is the solution of the Quasi-potential problem.

It is therefore the smoothest generalization to Quantum Field Theory of the Schroedinger equation. From the practical point of view, one has to make approximations to invert $T_1^J - T_2^J$.

The simplest approximation is known under the name of scalar P. A. If $|i\rangle$ is the initial state and $|f\rangle$ the final state, one replaces:

$$\langle f | T_1^J [T_1^J - T_2^J]^{-1} T_1^J | i \rangle \tag{3-d-1}$$

by

$$\langle f | T_1^J | i \rangle \frac{1}{\langle f | T_1^J - T_2^J | i \rangle} \langle f | T_1^J | i \rangle \tag{3-4-2}$$

A slightly more sophisticated approximation consists in splitting the two-body Hilbert space $H_2^J$ into the tensor product of the momentum part (w,p) and the spin part:

$$H_2^J = H_2^J\ (w,p) \otimes H_2^J\ (\text{spin}) \tag{3-4-3}$$

The space $H_2^J$ (spin) is always finite dimensional, and therefore one treats the inversion of $T_1^J - T_2^J$ exactly in $H_2^J$ (spin), while the approximation (3-d-2) is still made for the continuous part. This approximation is known under the name of Matrix Padé.

In fact, calculations with the Bethe-Salpeter equations (Gammel and Menzel, 1972) show that for the N-N system where $H_2$(spin) is a 16 dimensional space the [2/2] Matrix Padé is practically the exact solution, while at least a [4/4] scalar Padé is required to get the same precision. The Matrix Padé has also a deep physical meaning in the case of N-N scattering, linked with the pseudo-scalar nature of the force (pion) which couples strongly with the off-shell element ($\gamma_5$ is anti-diagonal).

Finally, we want to point out that in the case of the Schroedinger equation one can as well use all the time the [1/1] P. A., but include more and more off-shell elements to reach the exact solution, or use the on-shell P. A. approximant on the matrix element (scalar P. A.). It is then necessary to go to [2/2] or [3/3] P. A. to obtain a correct numerical convergence. In Field Theory, it seems easier to compute off-shell matrix [1/1] Padé, rather than [2/2] on-shell matrix Padé. However both must be done to allow comparison.

## 3-e. Models for which the convergence is proven rigorously

Up to now, we have only discussed practical or numerical convergence; it is interesting to study models for which we have a rigorous proof. The first models are to be found in potential scattering for regular potentials. For the partial waves, convergence has been proved for the T matrix (Chisholm, 1963;

Masson, 1967), and moreover the denominators of the P. A. have for limit the Jost functions (Garibotti and Villani, 1969). The difficulties found at any finite order, for the search of bound states due to the left hand cut singularity, were overcome using the off-shell P. A., starting from their variational principle generation (Alabiso et al., 1971). The results are very impressive and show that with the very low scalar P. A. [1/1] or [1/2], one gets the bound state with a few percent accuracy, even when the coupling constant reaches values of order of many tens. Numerical computations confirming the rigorous proofs have been performed by various authors for exponential and Yukawa potentials (Caser et al., 1969; Basdevant and Lee, 1969). The case of singular potentials with no change of sign or one change of sign produce a perturbation series with infinite Taylor coefficient; it has been treated by the use of cut-offs and the P. A. have been proved to have stationary values with respect to the cut-off, which tends to the exact solution (Garibotti et al., 1970; Fogli et al., 1971).

The Bethe-Salpeter equation in the ladder approximation in the scalar case has been proved, for the bound state, to have P. A. converging to the exact solution (Alabiso et al., 1972; Nieland and Tjon, 1970).

The Peres model and the vacuum polarization by a constant electric or magnetic field are other exactly solvable models for which the P. A. can be proved to converge (Graffi, 1971).

A very interesting model is provided by the anharmonic oscillator with an interaction

$$V(x) = P_{2n}(x) \tag{3-e-6}$$

where $P_{2n}(x)$ is a polynomial of degree 2n.

The perturbation series for the spectrum is of zero radius of convergence. If $n \leqslant 3$, then the P. A. can be shown to converge to the solution. If $n \geqslant 4$ then the Carleman criterion is violated, and the [N+J,N] P. A. converge to analytic functions; however, it cannot be proved that they have the same limit and that this limit is the solution. However the [N/N] and [N+1/N] still provide upper and lower bounds for the solution in this case (Loeffel et al., 1969; Simon, 1970; Graffi et al., 1970; Loeffel and Martin, 1970).

## 4. The strong interaction between pions and nucleons and the search for a Lagrangian

The first trials to use P. A. in field theory were done by using the $\lambda(\vec{\pi}\vec{\pi})^2$ model for evident reasons of simplicity. This model has the interesting feature that high order perturbation terms can be computed (Copley and Masson, 1967; Bessis and Pusterla, 1968).

In fact, for the s wave the [1/1], [1/2] and [2/1] scalar P. A. are found to differ by less than 7% in a rectangle:

$$4\mu^2 \leq \text{Res} \leq 100\mu^2 \quad ; \quad -100\mu^{2*} < \text{Ims} \leq 100^{\ 2}$$

However the $\sigma$ resonance is not found. For the p and d waves, the [1/2], [2/1] and [2/2] scalar P. A. have been computed and compared on the real s axis for $\text{Res} > 4\mu^2$. They differ by less than 10%. This time the $\rho$ and $f_0$ are correctly found, but an unwanted $I = 2$ resonance also appears degenerate with the $f_0$. The Regge intercept at $t = 0$ of the $\rho$ and $f_0$ as well as the slopes agree with experiment.

A more sophisticated model (Basdevant et al., 1969) including kaons gives correctly the two-body meson resonances with only 3 parameters:

$$g(\vec{\pi}\vec{\pi})^2 + \lambda(\vec{\pi}\vec{\pi})(\bar{K}K) + \nu(\bar{K}K)^2$$

However, although the I = 2 resonance of the previous model splits now from the $f_0$ in a nice way, yet unwanted degeneracies appear.

Models fulfilling current algebra have been investigated. The first one is the linear $\sigma$ model with a [1/1] scalar P. A. The s waves are now correct, as well as the p and d waves; however there is still an unwanted $I = 2$ resonance which is too low in energy (Basdevant and Lee, 1969).

A model with Yang-Mills fields (Basdevant and Zinn-Justin, 1972) presents the advantage of producing correct s and d wave

---

* When Ims is < 0 this corresponds to the second sheet values of the amplitude.

starting from the $\rho$. However this model does not reproduce observed effects when generalized to include the kaons (Iagolnitzer et al., 1972). Finally, the non-linear $\sigma$ model has not yet produced its best results due to the difficulties of renormalization.

For $\pi N$ scattering a first simple model using the Yukawa coupling $\mathcal{L}_I = g\bar{N}\,\gamma_5\,\vec{t}\,N\,\vec{\pi}$ gives rather poor results; but some waves are already good or very good (Mignaco et al., 1969; Mignaco and Remiddi, 1971).

By using the dispersive technique explained above, it has been possible with an input reduced to the nucleon pole plus a constant to reproduce all the physical results of $\pi N$ up to 1800 Mev with a [1/1] scalar P. A. (Fil'kov and Palyushev, 1972). This remarkable result allows us to think that the non-linear $\sigma$ model could be the key for the unique Lagrangian to produce $\pi\pi$, $\pi N$ and N-N scattering: because the input used for $\pi N$ is not essentially different from what the non-linear $\sigma$ model will produce if we knew how to renormalize it.

Finally, in the case of N-N scattering a first attempt was made with the Yukawa interaction and a scalar Padé (Bessis et al., 1970; Wortman, 1970), a second one has been done with the sub-matrix Padé in the non-linear $\sigma$ model as well as with the Yukawa model (Bessis et al., 1971). All features of the N-N scattering have been reproduced up to 400 Mev, with the exception of the two s-wave phase shift inversions. It is likely that with the full $16 \times 16$ Padé matrix those last features will also be reproduced.

## 5. Conclusion

In our opinion, there are two directions in which Padé approximants must now be applied:

- in axiomatic field theory, by trying to generalize the Villani theorem (Villani, 1972) for Stieltjes-Taylor series with infinite coefficients, to Hilbert space. One could then hope to prove an existence theorem for the $\lambda\phi^4$ theory, for which, due to the positivity of the interaction, we may expect to be in a generalized Stieltjes case.
- in phenomenology, by trying to define the unique Lagrangian which will be able to reproduce simultaneously the $\pi\pi$, $\pi N$ and N-N phase shifts by a low order Padé approximant in the low energy region (up to 1.5 Bev). It seems likely that the non-linear model, once we know how to renormalize it, will be the answer.

We are aware of the fact that many interesting topics in connection with Field theory have not been discussed. Among such, we quote the work of A. K. Common in 1969, who has applied Padé approximants to the Legendre series, and the work of Fleischer (1971) on the same type of argument. Crater (1970) has introduced an elegant method to solve the zero mass difficulties. Lopez and Yndurain in 1972 initiated a new and deep method of applying Padé approximants to the kaon-proton amplitudes, making use of the positivity; Yndurain (1972) also showed the importance of convergence in mean rather than uniform convergence. Finally, Professor M. Froissart has been at the origin of most of the papers produced by the Saclay group and must be specially quoted for his invaluable help.

Acknowledgments

I want to thank Professor R. Chisholm for his invitation to the Padé Conference in Canterbury. I also want to thank Dr. J. L. Basdevant, M. Pusterla and G. Turchetti for the fruitful discussions I have enjoyed. During a short stay at the Los Alamos Scientific Laboratory I have enjoyed warm hospitality from Dr. C. L. Critchfield and Dr. J. L. Gammel, and I wish to thank them for the assistance of their secretarial staff in the final typing of this paper.

References

Alabiso, C., Butera, P., Prosperi, G. M. (1971). Milano preprint

Alabiso, C., Butera, P., Prosperi, G. M. (1972). Milano preprint.

Basdevant, J. L., and Lee, B. W. (1969). Nuclear Phys. B13, 182.

Basdevant, J. L. and Lee, B. W. (1969). Phys. Letters 29B, 437.

Basdevant, J. L., Bessis, D., Zinn-Justin, J. (1969). Nuovo Cimento 60A, 185.

Basdevant, J. L. and Zinn-Justin, J. (1972). To be published in Physical Review.

Bessis, D. and Pusterla, M. (1968). Nuovo Cimento 54A, 243.

Bessis, D., Graffi, S., Grecchi, V. and Turchetti, G. (1970). Phys. Rev. D1, 2064.

Bessis, D. (1972). In the "Padé Summer School," Canterbury, England

Bessis, D. and Talman, J. D. (1972). In "The Boulder International Conference on Padé Approximants," to be published in the Rocky Mountain Journal of Mathematics.

Bessis, D., Turchetti, G. and Wortman, W. R. (1972). Physics Letters 39B, 601.

Bessis, D., Turchetti, G. and Wortman, W. R. (1972). In preparation.

Caneschi, L. and Jengo, R. (1969). Nuovo Cimento 60A, 25.

Caser, S., Piquet, C. and Vermeulen, J. L. (1969). Nuclear Phys. B14 119.

Chisholm, J. S. R. (1963). J. Math. Phys. 4, 1506.

Cini, M. and Fubini, S. (1954). Nuovo Cimento 11, 142.

Common, A. K. (1969). Nuovo Cimento 63A, 863.

Copley, L. A. and Masson, D. (1967). Phys. Rev. 164, 2059.

Crater, H. (1970). Phys. Rev. 2D, 1060.

Fil'kov, L. V. and Palyushev, B. B. (1972). Nucl. Phys. B42, 541.

Fleischer, J. (1971). Karlsruhe Preprint.

Fogli, G., Pellicoro, M. F. and Villani, M. (1971). Nuovo Cimento 6A, 79.

Gammel, J. L. and MacDonald, F. A. (1966). Phys. Rev. 142, 1145.

Gammel, J. L., and Menzel, M. T. (1972). Los Alamos Preprint LA-DC-72-511.

Garibotti, C. R. and Villani, M. (1969). Nuovo Cimento 61A, 747.

Garibotti, G. R., Pellicoro, M. V. and Villani, M. (1970). Nuovo Cimento 66A, 749.

Graffi, S., Grecchi, V. and Simon, B. (1970). Phys. Letters 32B, 631.

Graffi, S. (1971). Preprint, New York University 25/71.

Hurst, C. A. (1952). Phys. Rev. 85, 920.

Iagolnitzer, D., Zinn-Justin, J., Zuber, J. (1972). In preparation.

Kohn, W. (1947). Phys. Rev. 71, 635.

Lippmann, B. A. and Schwinger, J., (1950). Phys. Rev. 79, 469.

Loeffel, J. J., Martin, A., Simon, B. and Wightman, A. S. (1969). Phys. Letters 30B, 656.

Loeffel, J. J. and Martin, A. (1970). In "Cargese Lectures in Physics," (D. Bessis, editor), Gordon and Breach.

Lopez, C. and Yndurain, F. J. (1972). CERN Preprint TH-1511.

Masson, D. (1967). J. Math. Phys. 8, 2308.

Mignaco, J. A., Pusterla, M. and Remiddi, E. (1969). Nuovo Cimento 64A, 733.

Mignaco, J. A. and Remiddi, E. (1971). Nuovo Cimento 1A, 395.

Nieland, H. M. and Tjon, J. A. (1968). Phys. Lett. 27B, 309.

Nuttall, J. (1970). In "The Padé Approximant in Theoretical Physics" (G. A. Baker, Jr. and J. L. Gammel, eds.), Chapter 8, Academic Press, N. Y.

Peres, A. (1963). J. Math. Phys. 4, 332.

Simon, B. (1970). Ann. Phys. 58, 76.

Villani, M. (1972). In "Cargese Lectures in Physics" (D. Bessis, editor), Gordon and Breach.

Wortman, W. R. (1970). In "The Padé Approximant in Theoretical Physics" (G. A. Baker, Jr. and J. L. Gammel, eds.), Chapter 14, Academic Press. N. Y.

Yndurain, F. J. (1972). CERN Preprint TH-1521.

# MODEL FIELD THEORIES AND PADE APPROXIMANTS

M. Pusterla

(Istituto di Fisica dell'Università, Padova, Italy)

## 1. General observations

Padé approximants, once their convergence is under control, turn out to be very useful in the inspection of the physical content of models of field theory. More precisely, one looks at those cases, which are apparently the great majority of field theoretic models, where exact solutions are not known. Further, the strong coupling constants do not permit the assignment of any direct interpretation of the perturbative expansions of the physically relevant quantities.

There are many reasons why we are interested in analysing model field theories. We mention here several of them:

(a) the possibility of seeing in detail whether local renormalisable models admit solutions;

(b) the kind of physics one can develop from local Wick polynomials, chosen as interaction terms.

Both questions are extremely important: the first one is a fundamental point in axiomatic field theory; the second represents an answer, to a certain extent, to the open question of whether local Wick polynomials are meaningful and useful in the description of the interactions of elementary particle physics, beyond the domain of electrodynamics.

The simultaneous importance of (a) and (b) justifies the treatment of Lagrangians consisting of local fields with the addition of the typical features of the physical particles to which one refers, namely the spin, the isospin, etc.. In these notes I am considering in some detail the pion pion and the pion nucleon interactions and I mention only briefly the NN problem which will be discussed by Dr. Turchetti in the following paper.

## 2. The $\lambda\phi^4$ model

The physical realisation we consider here is the one where $\underline{\pi}$ represents a pion and must then be considered as a vector in isospin space. We allow direct forces among the pions without the introduction of other specific particles.

Our Lagrangian density is

$$= \tfrac{1}{2}\left[(\partial_\mu \underline{\pi})^2 - \mu_\pi^2(\underline{\pi})^2\right] - \tfrac{g}{4}(\underline{\pi}.\underline{\pi})^2 .$$

One is interested in the pion pion elastic scattering, at low and intermediate energy values (from threshold up to a total c.m. energy of 1.5 GeV approximately), in the bosonic resonances and in calculating the Regge trajectories. The essential ingredient to build up the approximate solution of the field theoretic model is the renormalised S matrix (or T matrix) expansion which has been computed by Copley and Masson (1967) up to third order and by Bessis and Pusterla (1968) up to fourth order.

The advantage of computing the fourth order is the ability to evaluate the [2,2], [1,3] and [3,1] approximants. The calculation of Bessis and Pusterla was performed with dispersion relation techniques and unitarity, so that it was possible to perform an analytical continuation into complex angular momentum (J), by the Froissart Gribov formula. The relativistic S matrix between the initial two pion state $|\alpha p_1, \beta p_2\rangle$ and the final state $|\gamma p_3, \delta p_4\rangle$, where $p_1, p_2, p_3, p_4$ are the four momenta and $\alpha, \beta, \gamma, \delta$ the isospin indices, can be written

$$S_{fi} = \delta_{fi} - i(2\pi)^{-2} \frac{1}{\sqrt{16\omega_1\omega_2\omega_3\omega_4}} \delta^{(4)} (P_i - P_f).$$

$$.\left[A\ \delta_{\alpha\beta}\delta_{\gamma\delta} + B\ \delta_{\alpha\gamma}\delta_{\beta\delta} + C\ \delta_{\alpha\delta}\delta_{\beta\gamma}\right],$$

where $s = -(p_1+p_2)^2$, $t = -(p_1-p_3)^2$, $u = -(p_1-p_4)^2$ and the crossing relations are

$$A(s,t,u) = A(s,u,t)$$
$$B(s,t,u) = C(s,u,t)$$
$$A(s,t,u) = B(t,s,u)$$
$$C(s,t,u) = C(t,s,u)$$
$$A(s,t,u) = C(u,t,s)$$
$$B(s,t,u) = B(u,t,s) \ ,$$

with the following projection operators (s channel)

$$P_0 = \frac{1}{3}\,\delta_{\alpha\beta}\delta_{\gamma\delta}$$
$$P_1 = \frac{1}{2}(\delta_{\alpha\gamma}\delta_{\beta\delta}-\delta_{\alpha\delta}\delta_{\beta\gamma})$$
$$P_2 = \frac{1}{2}(\delta_{\alpha\gamma}\delta_{\beta\delta}+\delta_{\alpha\delta}\delta_{\beta\gamma}) - \frac{1}{3}\,\delta_{\alpha\beta}\delta_{\gamma\delta}$$

and crossing matrix C

$$C = \begin{pmatrix} 1/3 & 1 & 5/3 \\ 1/3 & 1/2 & -5/6 \\ 1/3 & -1/2 & 1/6 \end{pmatrix}.$$

Let us call $\gamma^T(s,t,u)$, where T=0,1,2, the isospin amplitudes for the s channel; then the S matrix partial wave elements are, for all J values for which the following Froissart Gribov formula is convergent

$$S^T(J,K) = 1 - \frac{i}{16\pi^2 K\sqrt{K^2+\mu^2}} \int_{s_0}^{+\infty} ds' \mathrm{Abs}\ \gamma^T(s,s')\ Q_J(1+\frac{s'}{2K^2})$$

where $K^2 = \frac{1}{4}(s-4\mu^2)$, $s_0 \equiv 4\mu^2$ and Abs $\gamma(s,s')$ is the absorptive part of $\gamma$ for s fixed and $s'>s_0$.

Now, in order to compute the Padé approximants we need the perturbative (renormalised) absorptive parts $\gamma^T_{(n)}$, n being the order, which are computed from Feynman graphs and unitarity. One notices that the whole theory is generated by the renormalised coupling g and the unitarity condition, used perturbatively, and the formula is

$$\text{Abs}\ \gamma = -\frac{1}{2!}\sum_{(n)}^{el}(2\pi)^4\ \delta(P_i-P_n)\ \gamma^*_{ni}\gamma_{nf} +$$

$$+\frac{-1}{4!}\sum_{(n)}^{(in)}(2\pi)^4\ \delta(P_i-P_n)\ \gamma^*_{ni}\gamma_{nf}\ ,$$

where we have separated the two pion elastic intermediate states from the four pion inelastic ones. Two observations, to complete the scheme are necessary:

(a) $S_{(1)}$ is zero except for J=0 (this makes the $[1,1]$ approximant for partial waves valid only for S waves);

(b) the fourth order non-coplanar graph (four intermediate pions, in all channels) is absent in the T=1 amplitude.

The model gives the following results:

(a) when the isospin T=1, we obtain a resonance pole in the P wave amplitude $\pi\pi\to\pi\pi$ (a second sheet pole) which can be identified with the $\rho$ meson (for a value of g around 6);

(b) for the isospins T=0,2, we find also resonance poles in the D waves around the energy of the physical particle $f_0$ (more precisely at about 1500 MeV, instead of the experimental 1250 MeV); hence the model gives an exotic resonance in the T=2 D wave;

(c) the widths of $\rho$ and $f_0$ turn out narrower than the experimental values;

(d) S waves, both for the T=0 and T=2 isospin states, give negative phase-shifts and scattering lengths

$$a_o = -0.6m_\pi^{-1} \text{ and } a_2 = -0.3m_\pi^{-1}$$

whereas $a_o \simeq 0.18m_\pi^{-1}$ and $a_2 \simeq -0.05m_\pi^{-1}$ experimentally;

Furthermore, the S wave amplitudes are disconnected from the Carlsonian interpolation, e.g. vertex graphs in even signature amplitudes;

(e) the convergence of the [1,2], [2,1], [2,2], [1,3] and [3,1] Padé approximants for $J\neq 0$ have been tested from the Bessis Pusterla formulae for a wide range around the physical values of s and turns out to be excellent (within a few per cent); also for the S wave we have some estimate of convergence from [1,1] by a poloid test of crossing. This shows that the violation of crossing is very little around the physical region;

(f) the T=1 partial wave amplitudes allows an interpolation to continuous complex J values (odd signature) that gives the $\rho$ Regge trajectory with the correct intercept at s=0 and the experimental slope; the T=0, T=2 partial wave amplitudes do also interpolate giving two Regge trajectories (even signature): the one passing through the T=0 D wave pole ($f_o$) has experimental evidence and the calculated intercept and slope fall very near the experimental values; the T=2 trajectory is exotic.

## 3. Other $\pi\pi\rightarrow\pi\pi$ models

The so called $\sigma$ model was introduced by Gell-Mann and Levy (1960), and was used within the dynamical Padé programme by Basdevant and Lee (1969). The Lagrangian density is taken to be

$$\mathcal{L} = \frac{1}{2}\left[(\partial_\mu \underline{\pi})^2 + (\partial_\mu \sigma)^2\right] - \frac{\mu^2}{2}(\underline{\pi}^2+\sigma^2) - \frac{g}{4}(\underline{\pi}^2+\sigma^2)^2 + C\sigma.$$

If $C \equiv 0$ the theory treats the $\sigma$ as the fourth component of the four vector $(\pi_1, \pi_2, \pi_3, \sigma)$ in an abstract space and is chiral $SU(2) \otimes SU(2)$ invariant. For $C \neq 0$ the chiral symmetry is broken and the P.C.A.C. condition $\partial^\mu \underline{A}_\mu = C\underline{\pi}(x)$ is valid, $\underline{A}_\mu$ being the axial vector current. The Lagrangian gives tree diagrams that provide the results of the current algebra in the soft pion limit. The theory is renormalisable in such a way to preserve the current algebra constraints and P.C.A.C. in each perturbative order. The parameter C is simply related with the lifetime of the pion: $C = f_\pi m_\pi^2$. The parameters of the theory are $\mu^2$, C and g alternatively, $m_\pi^2$, $f_\pi$ and g.

Only the [1,1] approximant was computed and the results seem very promising:

(a) the $\pi\pi$ S waves are much more realistic; T=0 is attractive, T=2 is repulsive, a T=0 S wave resonance occurs at $M_\sigma \simeq 500$ MeV, $\Gamma_\sigma \simeq 300$ MeV, $a_0 \simeq 0.15 m_\pi^{-1}$, $a_2 \simeq -0.04 m_\pi^{-1}$;

(b) almost same predictions are found as for the $\lambda\phi^4$ model for P waves and D waves. A better position of the $f_0$ particle is obtained and the exotic resonance in the T=2 D wave still is present but pushed to higher values of the mass: more precisely $m_\rho \simeq 780$ MeV, $\Gamma_\rho \simeq 35$ MeV, $M_{f_0} \simeq 1280$ MeV and $\Gamma_{f_0} \simeq 180$ MeV.

## 4. The Yang Mills field

The Yang Mills $\rho$ field has been looked at by Basdevant and Zinn-Justin (1971) by using the Lagrangian density:

$$\mathcal{L} = \frac{1}{2}(\partial_\mu \underline{\pi} - g\underline{\rho}_\mu \times \underline{\pi})^2 - \frac{1}{2} m_\pi^2 (\underline{\pi}.\underline{\pi})$$

$$- \frac{1}{4}(\partial_\mu \underline{\rho}_\nu - \partial_\nu \underline{\rho}_\mu - g\underline{\rho}_\mu \times \underline{\rho}_\nu)^2 - \frac{1}{2} m_\rho^2 (\underline{\rho}_\mu . \underline{\rho}^\mu),$$

when $\underline{\rho}_\mu$ is an isovector vector field ($\rho$ meson).

From the $[1,1]$ approximant the authors obtain strong S wave attraction in T=0 and a broad $\sigma$ resonance ($M_\sigma \simeq 450$ MeV, $\Gamma_\sigma \simeq 500$ MeV), $a_0 \simeq 0.17 m_\pi^{-1}$, $a_2 \simeq -0.06 m_\pi^{-1}$. The $\rho$ meson is obviously at the correct mass and width ($m_\rho$ and g) and the $f_0$ at $M_{f_0} \simeq 1280$ MeV, $\Gamma_{f_0} \simeq 250$ MeV. There are no T=2 exotic resonances.

## 5. Pion nucleon interactions

The pion nucleon interaction is probably the problem, in elementary particle physics, where theory and experiment can be compared more directly throughout the whole energy range of scattering which goes from the threshold up to 2 GeV c.m. total energy. This comparison has become possible because of the accurate experimental data available for the elastic scattering which allow the phase-shift analysis.

It is worth remembering the main points of the relativistic kinematical formalism for the $\pi N$ elastic $S_{fi}$ matrix elements

$$S_{fi} = \delta_{fi} - i(2\pi)^{-4}\, \delta^{(4)}(P_f - P_i)\, \frac{m}{\sqrt{4\omega_1\omega_2 E_1 E_2}}\, \bar{u}_f\, \mathrm{T} u_i$$

$$\mathrm{T} = -A(s,t,u) + i\not{Q}B(s,t,u)$$

$$A = \delta_{\alpha\beta}A^{(+)} + \tfrac{1}{2}[\tau_\beta,\tau_\alpha]\, A^{(-)} \qquad A^{(+)} = \tfrac{1}{3}[A^{\frac{1}{2}} + 2A^{3/2}]$$

$$B = \delta_{\alpha\beta}B^{(+)} + \tfrac{1}{2}[\tau_\beta,\tau_\alpha]\, B^{(-)} \qquad A^{(-)} = \tfrac{1}{3}[A^{\frac{1}{2}} - A^{3/2}]$$

$$A^{(\pm)}(s,t,u) = \pm A^{(\pm)}(u,t,s) \qquad u = -(p'-q)^2$$

$$B^{(\pm)}(s,t,u) = \mp B^{(\pm)}(u,t,s) \qquad t = -(p'-p)^2$$

$$s = -(p+q)^2 \ .$$

The partial wave S matrix elements $S^T_{\ell\pm}(W)$ are computed with the following formula due to McDowell

$$\eta^T_{\ell\pm}(W)e^{2i\delta^T_{\ell\pm}(W)} \equiv S^T_{\ell\pm}(W) = 1 + 2iK \frac{1}{8\pi W}\{(E+m)\left[A^T_\ell+(W-m)\ B^T_\ell\right]$$

$$+ (E-m)\left[-A^T_{\ell\pm 1}+(W+m)\ B^T_{\ell\pm 1}\right]\} .$$

Whenever one is given a regular field theory, each order and each graph at a given order provide its contribution to the $S^{T(n)}_{\ell\pm}$. The actual computation of the $A'_s$ and the $B'_s$ is achieved from Feynman rules or dispersion relations and unitarity. The [1,1] approximant has been applied by Mignaco, Pusterla and Remiddi (1969) and independently by Gammel and collaborators to the Lagrangian

$$\mathcal{L}_{int} = -ig\bar{N}\ \gamma^5\ \underline{\tau}N\cdot\underline{\pi} - \frac{\lambda}{4}\ (\underline{\pi}\cdot\underline{\pi})^2$$

and by Mignaco and Remiddi (1971) to the σ model

$$\mathcal{L}_{\sigma} = \bar{N}\left[i\partial_\mu\gamma^\mu - g(\sigma+i\underline{\pi}\cdot\underline{\tau}\gamma^5)\right]\ N+\frac{1}{2}\left[(\partial_\mu\sigma)^2+(\partial_\mu\underline{\pi})^2\right]+$$

$$+\ \frac{1}{2}\mu^2(\sigma^2+\underline{\pi}^2)\ -\ \frac{\lambda}{4}(\sigma^2+\underline{\pi}^2)\ +\ C\sigma\ .$$

The predictions of both models obtained from the [1,1] appear very inadequate to describe the physical situation. One, however, cannot decide whether this negative result has to be attributed to the fact that the [1,1] is too far away from the limit (very weak convergence) or to the physical content of the Lagrangians considered which may appear too poor. More specifically, one finds that the $\gamma^5$ theory does not reproduce the pion nucleon phase-shifts except in a very qualitative fashion for the isospin 3/2 waves, even if one plays around with the λ parameter. As far as the σ model is concerned, the Lagrangian gives the result of the current algebra in the tree approximation but the scattering lengths are obtained from the Born term only under the condition $\mu_\sigma > m_N > \mu_\pi$.

Such a condition makes the Born term inadequate to reproduce the scattering lengths even in sign: it gives the same signs instead of the experimental observation of opposite signs.

We finally come to the most spectacular results obtained in the pion nucleon problem, assuming that the calculations will be confirmed in the future: the Padé calculation worked out by Fil'Kov and Palyushev. These authors start from the $\pi N$ dispersion relations unsubtracted for $A^{(-)}$, $B^{(\pm)}$ and subtracted for $A^{(+)}$. They justify the subtraction of $A^{(+)}$ because the Pomeranchuk pole is present in this amplitude and absent in the $A^{(-)}$. Let us call $(s_o, t_o, u_o)$ the subtraction point. They have then

$$B^{(\pm)} = g^2\left(\frac{1}{m^2-s} \mp \frac{1}{m^2-u}\right) + \frac{t}{\pi}\int_{s_o}^{+\infty} ds' \, \mathrm{Im}_s\, B^{(\pm)}(s',t) \cdot$$

$$\cdot \left[\frac{1}{s'-s} \mp \frac{1}{s'-u}\right]$$

$$A^{(-)} = \frac{1}{\pi}\int_{s_o}^{+\infty} ds' \, \mathrm{Im}_s\, A^{(-)}(s',t)\left(\frac{1}{s'-s} - \frac{1}{s'-u}\right)$$

$$A^{(+)} = a + \frac{t}{\pi}\int_{t_o}^{+\infty} dt' \, \frac{\mathrm{Im}\, A^{(+)}(t',u_o)}{t'(t'-t)} - \frac{t}{\pi}\int_{s_o}^{\infty} ds' \, \frac{\mathrm{Im}_s\, A^{(+)}(s',u_o)}{(s'-s_o)(s'-s_o+t)}$$

$$+ \frac{u-u_o}{\pi}\int_{s_o}^{+\infty} ds' \, \mathrm{Im}_s\, A^{(+)}(s',t)\left[\frac{1}{(s'-u)(s'-u_o)} - \frac{1}{(s'-s_o)(s'-s_o+t)}\right]$$

The presence of the dispersive integral

$$\int_{t_o}^{+\infty} dt' \, \frac{\mathrm{Im}\, A^{(+)}(t',u_o)}{t'(t'-t)}$$

would impose a treatment of the annihilation channel $N\bar{N}\to 2\pi$ in the isospin T=0 but the authors avoid this difficulty by dealing with this term phenomenologically

$$\int_{t_o}^{+\infty} dt' \, \frac{\mathrm{Im}_t\, A^{(+)}(t',u_o)}{t'(t'-t)} \simeq \frac{\beta}{(m_\sigma^2-t)} \, .$$

The method of Fil'Kov and Palyushev consists of solving the dispersive equations written above by iteration; they use the unitarity condition which is a nonlinear relation and allows the evaluation of the imaginary part at higher orders in terms of the whole amplitude at lower orders. Their initial term for iteration (the Born term in Lagrangian language) is

$$B^{(\pm)} = g^2\left(\frac{1}{m^2-s} \mp \frac{1}{m^2-u}\right)$$

$$A^{(+)} = a + \frac{\beta t}{m_\sigma^2-t}$$

$$A^{(-)} = 0 \; .$$

They calculate only the first term that comes after the Born approximation and can compute then the partial wave approximants [1,1] for the waves of experimental interest (S, P and D). The constant a is evaluated in terms of the S and P wave scattering lengths

$$A^{(+)}(s_o,t_o,u_o) = a = 4\pi\{\frac{2m+\mu}{6m}(a_1+2a_3) - \frac{2}{3}m\mu \cdot$$

$$\cdot \left[(a_{11}+2a_{31}) - (a_{13}+2a_{33})\right]\} \; .$$

We now summarise and comment on the predictions of this calculation which, if confirmed, is very important for future developments in our opinion.

$S_3$ and $P_{31}$

The phase-shifts of the T=3/2 S wave ($S_3$) and J=1/2 P wave ($P_{31}$) coincide with the experimental ones from threshold up to 1400 MeV and to 1600 MeV respectively very accurately (negative and smoothly decreasing). Beyond these energy values and up to 1800 MeV the theoretical predictions are still quite good, but do not follow as accurately as before the experimental data.

This, however, may be attributed to the inelasticity effects that are not taken into account by the [1,1] Padé approximant. One should notice that also the previous models work nicely for these particular waves.

$P_{33}$

The $P_{33}$ is the resonating wave that many models reproduce from the times of the static Chew Low theory; of course, the calculation of Fil'Kov and Palyushev predicts accurately the experimental $P_{33}$ phase-shift from the threshold up to 1800 MeV (much above the 33 resonance).

$D_{33}$ and $D_{35}$

Both the T=3/2 D waves are qualitatively correct and up to 1500 MeV are quantitatively reproduced. The phase-shifts are both negative and decreasing.

Again the previous models, mentioned above, agree qualitatively with the data. It is, however, in the T=1/2 amplitudes that the previous attempts failed clamorously. The Fil'Kov Palyushev calculation gives very good results in these isospin states.

$S_1$

Very good predictions are here obtained in the $S_1$ (J=1/2, T=1/2 wave) when the phase-shift is positive and increasing, closely following the experimental behaviour, whereas the preceding models give completely wrong results.

$P_{11}$

For the $P_{11}$ (J=1/2, T=1/2) we have the most spectacular prediction of Fil'Kov and Palyushev: the famous $P_{11}$ (Roper) resonance is here predicted with great accuracy, besides the behaviour of this phase-shift before and after the resonance. It is to be noticed that all previous attempts of any kind to reproduce such a resonance failed.

$\underline{P_{13}}$

The $P_{13}$ is predicted very close to zero and in fact it is so experimentally: the experimental phase-shift is negative but very near to zero.

$\underline{D_{13} \text{ and } D_{15}}$

The two T=1/2 D waves are reproduced quantitatively up to 1500 MeV and qualitatively above this point. One must attribute the deviations, in particular above 1500 MeV, to the lack of inelasticity in the [1,1] approximant. All the results mentioned are obtained with one parameter $\beta \simeq 0.25$ ($m_\sigma$) but $\beta=0$ also satisfies the data reasonably well. It is interesting to point out that for the case $\beta=0$ the Born term of Fil'Kov and Palyushev can be derived from an effective Lagrangian ($\bar{N}N\underline{\pi}\cdot\underline{\pi}$) which appears in the non-linear $\sigma$ model.

## 6. Nucleon Nucleon interaction

Historically, the Yukawa (pseudoscalar) interaction was the first problem investigated. Quantitative calculations of the phase-shifts were performed independently by Wortman (1969) in America and Bessis, Graffi, Grecchi and Turchetti (1970) in Saclay and Bologna. The results of the first investigation with the [1,1] approximant were encouraging.

The introduction of the linear and non-linear $\sigma$ model showed some improvements in the S wave calculation while the $\gamma^5$ interaction appears to cover most of the dynamics at low energy. The greatest improvements are due to the use of the [1,1] matrix Padé approximant worked out in the spin space by Bessis and Turchetti, which, except for the S waves, improves the previous results and yields much closer agreement between predictions and the experimental phase-shifts. I refer to the contributed paper of Dr. Turchetti for details on the matrix method and the results.

References

Bardeen W. and Lee B. W., (1969). Phys. Rev. 177, 2389.

Basdevant J. L., Bessis D. and Zinn-Justin J., (1969). Nuovo Cim. 60A, 185.

Basdevant J. L. and Lee B. W., (1969). Phys. Letters 29B, 437.

Basdevant J. L. and Zinn-Justin J., (1971). Phys. Rev. D3, 1865.

Basdevant J. L. and Zinn-Justin J., (1969). Private and unpublished communication.

Bessis D. and Pusterla M., (1968). Nuovo Cim. 54A, 243.

Bessis D., Graffi S., Grecchi V. and Turchetti G., (1970). Phys. Rev. 1D, 2064.

Bessis D., Turchetti G. and Wortman W., (1972). Phys. Letters 39B, 601.

Copley L. and Masson D., (1967). Phys. Rev. 164, 2059.

Fil'Kov J. V. and Palyushev B. B., (1972). Nucl. Phys. B42, 541.

Gammel J. L., Menzel M. T. and Kubis J. J., (1968). Phys. Rev. 172, 1664.

Gell-Mann M. and Levy M., (1960). Nuovo Cim. 16, 705.

Mignaco J. A., Pusterla M. and Remiddi E., (1969). Nuovo Cim. 64A, 733.

Mignaco J. A. and Remiddi E., (1971). Nuovo Cim. 1A, 395.

Wortman W. R., (1969). Phys. Rev. 176, 1762.

# PADE APPROXIMANTS IN NUCLEON-NUCLEON DYNAMICS

G. Turchetti

(Istituto di Fisica della Università, Bologna, Italy)

In strong interactions it is often found that the second term in the perturbation expansion for the S matrix is much larger than the first one. This phenomenon has a very simple physical explanation: strongly interacting systems exhibit bound states and resonances which correspond to polar singularities in the coupling constant $\alpha$ and which cause the asymptotic expansion to diverge.

In order to extract the relevant information corresponding to the physical value of $\alpha$, we replace the perturbation series by a sequence of Padé approximants. For series of Stieltjes, the P.As. provide the correct analytic continuation in all the complex plane including poles, even when the radius of convergence shrinks to zero.

In scattering theory, though the S matrix cannot be proved in general to be a series of Stieltjes the use of P.A. is still supported by strong physical arguments: in fact the P.As. are unitary and are derived from variational principles.

## The operator Padé approximant

In field theory the $[N/N]$ P.As. to the Born series of the transition operator T are obtained using a very natural ansatz (Cini and Fubini, 1954) to the Lippmann-Schwinger (1950) variational principle. Given the S operator, T is defined by

$$S = I + iT \quad \text{and} \quad T = \alpha\, T_1 + \alpha^2\, T_2 + \dots .$$

Since our first aim is to treat low energy elastic processes we restrict ourselves to the Hilbert space of two particle states H, where the matrix elements of $T_n$ can be computed using Feynman rules and standard renormalisation.

Let us recall here the properties of the simplest approximant. The $[1/1]$ operator Padé approximant (O.P.A.) which is

$$[1/1]_{T(\alpha)} = \alpha\, T_1 [T_1 - \alpha\, T_2]^{-1}\, T_1 \qquad (1)$$

i) is the simplest solution of the Schwinger variational principle;

ii) fulfills the unitarity condition $T - T^+ = iTT^+$ in H;

iii) is the solution of the quasi-potential problem;

iv) when $T_2 = T_1\, G_0\, T_1$ is reducible, it is the exact solution of the Bethe-Salpeter equation $T = V + V\, G_0\, T$ with a potential $V = \alpha\, T_1$.

The invariance under space rotations allows one to diagonalise T with respect to the angular momentum J:

$$[1/1]_{T(J)}\ (\alpha) = \alpha\, T_1^{(J)} [T_1^{(J)} - \alpha\, T_2^{(J)}]^{-1}\, T_1^{(J)} \qquad (2)$$

and due to covariance of P.As. under similarity transformations, (2) can be deduced from (1).

As a consequence of the above properties the $[1/1]$ O.P.A. seems to be, assuming the existence of $(T_1 - \alpha\, T_2)^{-1}$, the smoothest and most consistent generalisation of the Schrödinger equation to field theory.

## The scalar and matrix Padé approximants

The evaluation of $[1/1]_{T(J)}$ between an initial state $|i\rangle$ and a final state $|f\rangle$ is a formidable task and some simpler approximations are required: the scalar Padé approximant (S.P.A.) replaces the operators $T_K^{(J)}$ by their matrix elements between $|i\rangle$ and $|f\rangle$. By definition the S.P.A. is the P.A. to the Taylor expansion of $\langle f|\ T^{(J)}\ (\alpha)\ |i\rangle$.

With spinning particles several amplitudes arise for fixed total angular momentum and the S.P.A. applies to each of them. However the most exhaustive treatment of spin is obtained through the matrix Padé approximant (M.P.A.) which is devised to extract all the information contained in the spin structure of the amplitude. In order to exhibit the connection between the O.P.A. and the M.P.A. we consider the explicit case of two spin $\frac{1}{2}$ particles.

The Hilbert space H of two free fermions is spanned by the vectors $|v\rangle = |\mu_1, p_1\rangle|\mu_2, p_2\rangle$ which belong to the irreducible representations of the Poincaré group: p is a continuous index labelling the momentum whereas μ is a discrete index labelling the $(\frac{1}{2},0) \oplus (0,\frac{1}{2})$ representation of $SU(2) \otimes SU(2)$ isomorphic to the Lorentz group. We decompose H into the direct product of a momentum space $H_p$ spanned by $|p_1, p_2\rangle$ and a spin space $H_s$ spanned by $|\mu_1, \mu_2\rangle$

$$H = H_p \otimes H_s \qquad |v\rangle = |\mu_1, \mu_2\rangle|p_1, p_1\rangle \ .$$

Let $|v\rangle$ and $|v'\rangle$ be two vectors corresponding to the initial and final momentum states and define

$$\langle v'|T^{(J)}|v\rangle = \langle T^{(J)}(p_1', p_2', p_1, p_2)\rangle_{\mu'\mu}$$

where $\mu \equiv (\mu_1, \mu_2)$, $\mu' \equiv (\mu_1', \mu_2')$ and $\langle T^{(J)}\rangle$ is a 16×16 matrix acting on $H_s$. The M.P.A. is obtained from the O.P.A. when we restrict ourselves to the spin space and replace the operators in $H_p$ by their matrix elements between the initial and final momentum states. The M.P.A. is therefore to be defined as the P.A. to the Taylor expansion of the matrix $\langle T^{(J)}(\alpha)\rangle$.

## The NN interaction

The NN system is very suitable to test the various P.A. previously proposed since the Lagrangian is known to some extent and the experimental situation very well defined. The problem has been investigated in potential scattering but we believe that due to the pseudoscalar character of the pion even at very low energy the nature of the interaction is relativistic and field theory is the most appropriate scheme to treat it. The basic interaction Lagrangian is the Yukawa model

$$L_y = - i\, G\, \tilde{\psi}\, \gamma_5\, \vec{\tau}.\vec{\pi}\, \psi - \frac{\lambda}{4} \vec{\pi}^4 .$$

The [1/1] S.P.A. which was first computed by Bessis et al. (1970) provided a set of results in good qualitative agreement with experiment. A bound state was found only in the channel with the deuteron quantum numbers and the phase shifts were a remarkable improvement over the unitarised Born approximation. The S.P.A. failures very likely originate in the singular structure of the Born term which contributes only to the pseudoscalar Fermi invariant. This invariant vanishes at threshold and consequently the S waves have zero scattering lengths and the threshold behaviour of a P wave. The matrix Padé approximant overcomes this difficulty since the Born term is not more singular. As the angular momentum diagonalisation is rather cumbersome in the general case, we have first computed the [1/1] M.P.A. to that particular 2×2 submatrix to which the 16×16 matrix exactly reduces at threshold (Bessis, Turchetti and Wortman 1972). The improvements are remarkable: all the phase shifts, except the S waves, exhibit a semiquantitative agreement with experiment. The S waves at very low energy are qualitatively correct, but the sign inversion does not yet occur.

The inclusion of strange particles proves to be irrelevant (Wortman 1969), while the chiral Lagrangians (which satisfy the current algebra constraints) improve the S waves at very low energy without producing the sign inversion (Turchetti et al. 1971, 1972). If the negative tail of the S waves is mainly due to one and two pion forces the complete [1/1] M.P.A. for

Yukawa or the non-linear σ model will be able to reproduce it and show that good convergence is almost achieved.

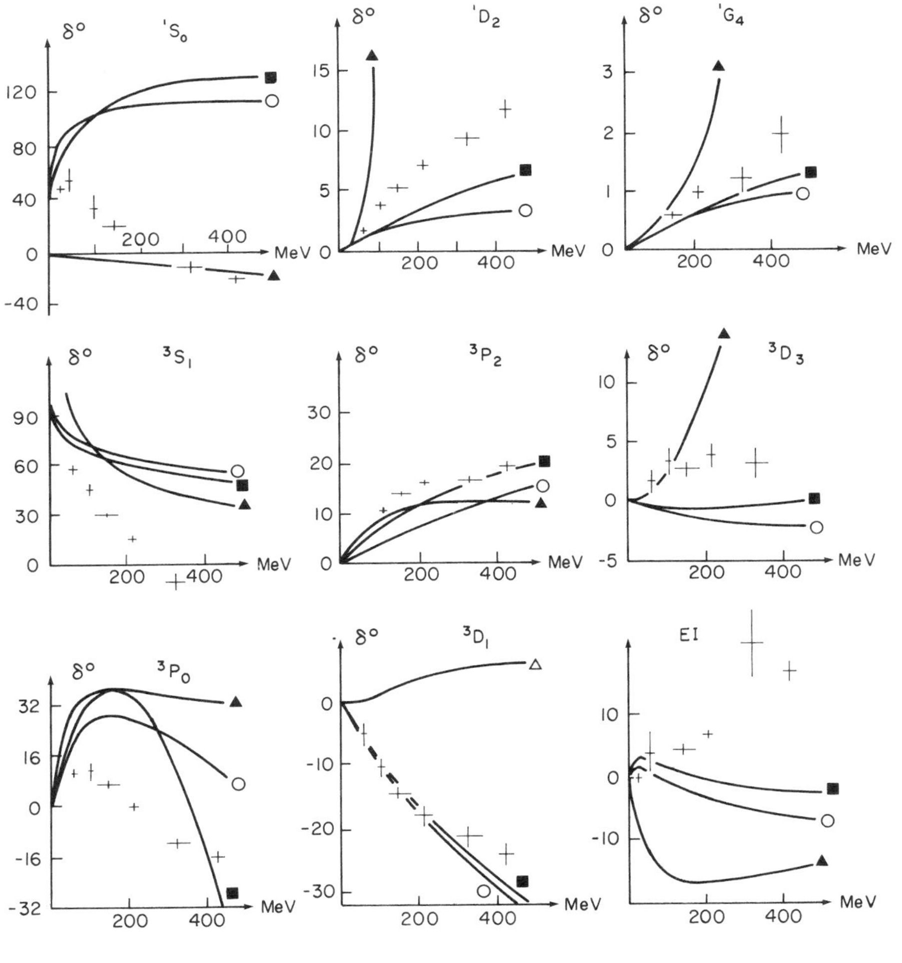


The figures show the data on phase shifts (in degrees) versus the nucleon kinetic energy (in MeV). The three continuous curves, labelled + for the Yukawa model with S.P.A., * for the Yukawa model with M.P.A. and O for the σ model with M.P.A., show the theoretical results using the [1/1] approximant.

Acknowledgment

It is a pleasure to thank Dr. D. Bessis for useful discussions and Professor J. Wm. McGowan for hospitality at the Western Ontario University.

References

Barlow R. and Bergère M., (1972). Saclay preprint DPhT. 71, 72.

Bessis D., Graffi S., Grecchi V., Turchetti G., (1970). Phys. Rev. 1, 2064.

Bessis D., Turchetti G., Wortman W., (1972). Phys. Lett. 39B, 601.

Cini M. and Fubini S., (1954). Nuovo Cimento 11, 142.

Graffi S., Grecchi V. and Turchetti G., (1972). Nuovo Cimento Lett. 4, 281.

Lippmann B. A. and Schwinger J., (1950). Phys. Rev. 79, 569.

Turchetti G., (1971). Proceedings of Marseille Symposium (A. Visconti, ed.) CNRS Marseille.

Wortman W. R., (1969). Phys. Rev. 176, 1762.

# SIMULATION AND CONTROL

# APPLICATIONS OF PADE APPROXIMANTS IN ELECTRICAL NETWORK PROBLEMS

M. I. Sobhy

(Electronics Laboratory, University of Kent, England)

## 1. Introduction

This paper reviews the various areas in the field of network analysis and synthesis in which continued fractions and Padé approximants are used.

In frequency domain problems, continued fractions are used to develop ladder networks. This is particularly successful in circuits containing only two types of circuit elements where a solution is always possible (Gillemin, 1957, Newcomb, 1966 and Weinberg, 1962). A continued fraction solution is not guaranteed in the case of circuits R-L-C circuits unless certain limitations are imposed. Predistortion techniques can be used to solve these networks (Anderson and Lee, 1966 and Darlington, 1939).

Padé approximants can be used to synthesise circuits for a specified time domain response (Teasdale, 1953 and Weinberg, 1962). A method is described by which control over the error in the time domain is achieved. A few introductory paragraphs are given in this paper to introduce the basic concepts of network problems.

## 2. The Circuit Elements

Passive electrical networks contain resistors, inductors and capacitors. The circuit elements could be connected together in any arbitrary configuration to form the desired network. Table I defines the relationships between the currents through the elements and the voltages across them.

| Element | Time Domain | Frequency Domain | Impedance Z | Admittance Y |
|---|---|---|---|---|
| Resistor, R | $v=Ri$ | $V=RI$ | $R$ | $\frac{1}{R}$ |
| Inductor, L | $v=L\frac{di}{dt}$ | $V=sLI-i(0+)L$ | $sL$ | $\frac{1}{sL}$ |
| Capacitor, C | $v=\frac{1}{C}\int idt$ | $V=\frac{I}{sC}+\frac{v(0+)}{s}$ | $\frac{1}{sC}$ | $sC$ |

R, L and C are positive and real quantities.

$s = \sigma+j\omega$ is the complex frequency.

Table I

In later discussions we will also deal with active networks which contain negative resistors.

## 3. Driving Point Impedance and Positive Real Functions

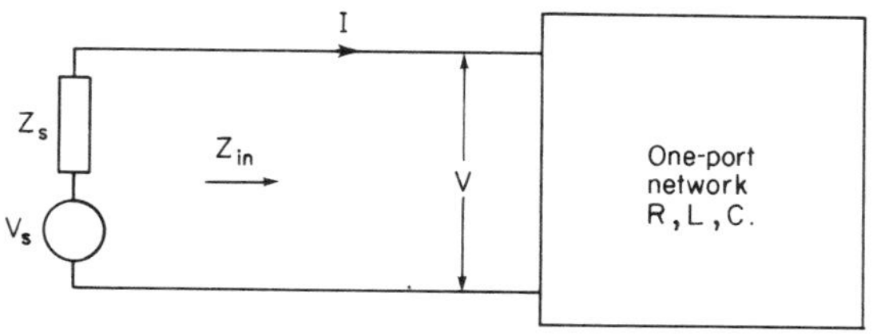


For a single port network the driving point or input impedance is that impedance seen at the input terminals by the source, $Z_{in} = V/I$.

Since $Z_{in}$ is a function of the impedances given in Table I and the way they are connected, it follows that if the complex frequency s is real then the driving point impedance $Z_{in}$ is also real. This is a necessary condition for the impedance function $Z_{in}$ to represent the input impedance of a passive network.

$P_{av}$ is the average power delivered to the network by the source. If the circuit is passive $P_{av}$ will always be positive,

$$P_{av} = \frac{1}{2} \operatorname{Re}(Z_{in}(j\omega))|I|^2 > 0 \quad . \qquad (1)$$

It follows from (1) that if the real part of s is positive then $P_{av}$ will always be positive if the real part of $Z_{in}(s)$ is positive.

We can now write the necessary and sufficient condition for an impedance function Z(s) to represent a passive network.

$$Z(s) \text{ is real if } s \text{ is real}$$

$$\operatorname{Re} Z(s) \geq 0 \text{ if } \operatorname{Re} s \geq 0 \quad . \qquad (2)$$

Functions that satisfy (2) are called positive real (p.r.) functions.

## 4. Continued Fraction Representations of Ladder Networks

Although the elements of a network can in general be connected together in any arbitrary manner, we are going to consider a special class of networks which can be directly represented by a continued fraction expansion. These are called ladder networks and contain alternate series impedances and shunt admittances. This class of networks has a wide range of applications in filter and equaliser circuits.

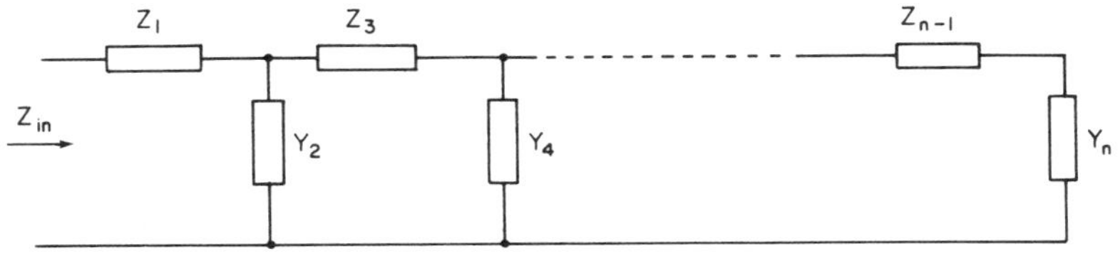


Fig. 2

The input impedance Z(s) of the ladder network shown in Fig. 2 is given by

$$Z(s) = Z_1 + \frac{1}{Y_2} + \frac{1}{Z_3} + \ldots \frac{1}{Z_{n-1}} + \frac{1}{Y_n} \ .$$

This means that if a continued fraction expression can be found for the given function Z(s) a network of the type shown in Fig. 2 will realise the required impedance. However, this is only true if each term of the continued fraction expansion is a positive real function and thus can be realised using passive circuit elements.

It is always possible to satisfy this condition if the given impedance represents a network containing only two circuit elements (L-C, R-L or R-C). This is because the poles and zeros of these functions are restricted in such a way to make all terms of the expansion realisable.

These restrictions are:

(a) For L-C circuits all the poles and zeros lie on the $j\omega$ axis, they alternate and there is always a pole or a zero at $s=0$ and at $s=\infty$.

(b) For R-L or R-C impedances all the poles at the zeros lie on the negative real axis and they alternate.

## 5. R-L-C Ladder Networks

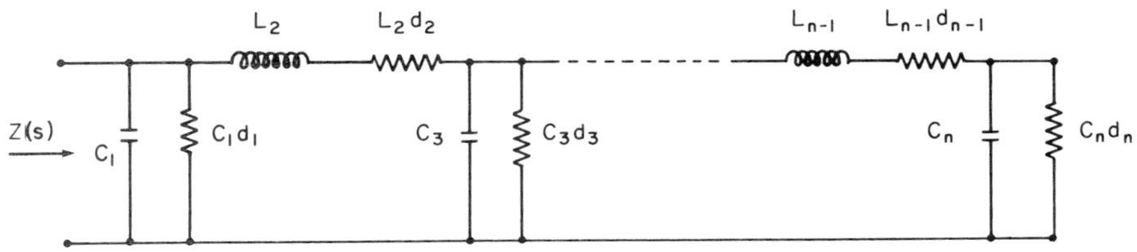


Fig. 3

Fig. 3 shows a lossy ladder network. The continued fraction form of the input impedance of this network is given by

$$Z(s) = \frac{1}{C_1(s+d_1)} + \frac{1}{L_2(s+d_2)} \cdots + \frac{1}{L_{n-1}(s+d_{n-1})} + \frac{1}{C_n(s+d_n)} \,. \tag{3}$$

Ideally there should be no restriction in the values of the loss factors $d_1, d_2, \ldots, d_n$. However, it is not easy to find the conditions imposed on the impedance function Z(s) so that the continued fraction form of the form given in (3) can be obtained.

Later it will be shown that with the appropriate restrictions on the loss factors an expansion can always be found.

## 6. Transfer Functions

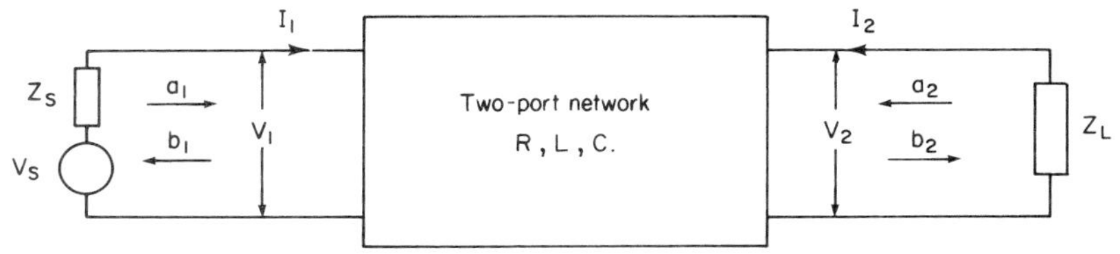


Fig. 4

Transfer functions describe the characteristics of two port networks. They relate the output quantities of the network (either the voltage, the current or the incident and reflected quantities $a_1$ and $b_1$) to the input quantities. The network is represented by a 2×2 matrix relating these quantities.

The impedance matrix is defined by

$$\begin{pmatrix} V_1 \\ V_2 \end{pmatrix} = \begin{pmatrix} Z_{11} & Z_{12} \\ Z_{21} & Z_{22} \end{pmatrix} \begin{pmatrix} I_1 \\ I_2 \end{pmatrix} .$$

The admittance matrix is the inverse of the impedance matrix. Alternatively we can represent the network by the scattering matrix

$$\begin{pmatrix} b_1 \\ b_2 \end{pmatrix} = \begin{pmatrix} s_{11} & s_{12} \\ s_{21} & s_{22} \end{pmatrix} \begin{pmatrix} a_1 \\ a_2 \end{pmatrix} .$$

The impedance matrix Z is related to the scattering matrix S by

$$(S) = (Z + I_n)^{-1}(Z - I_n)$$

where $I_n$ is a unit matrix.

For a passive network the impedance matrix Z is a positive real matrix which means that

$$Z(s) \text{ is holomorphic in } \sigma > 0$$
$$Z^*(s) = Z(s^*) \text{ in } \sigma > 0$$
$$Z_H(s) > 0 \text{ in } \sigma > 0 \quad .$$

In these equations, we use the convention that * denotes complex conjugate, ∿ denotes transpose and $Z_H(s)$ is the Hermitian part of $Z(s)$, namely $\frac{1}{2}(Z(s)+\tilde{Z}^*(s))$.

From these conditions imposed on Z similar conditions are derived for the scattering matrix S, and these are

$$S(s) \text{ is holomorphic in } \sigma > 0$$

$$S^*(s) = S(s^*) \text{ in } \sigma > 0$$

$$I_n - \tilde{S}^*(s)\, S(s) > 0 \text{ in } \sigma > 0 .$$

Furthermore, for a lossless circuit

$$Z = -\tilde{Z}_* \ , \quad Z(s) = -\tilde{Z}(-s) \quad \text{and} \quad S = \tilde{S}_*^{-1}$$

where

$$Z_*(s) = Z(-s) \quad \text{and} \quad S_*(s) = S(-s) .$$

It has been shown that (Newcomb, 1966) the above conditions can be satisfied if the matrix S is in the form

$$S = \frac{1}{d}\begin{pmatrix} h & f \\ f & \mp h_* \end{pmatrix} .$$

Where f is either an even or an odd function of s, the negative sign is used when f is even and the positive sign when it is odd, and d is a Hurwitz polynomial. Thus for a lossless circuit $S_{21}$ is given by $S_{21} = f/d$, and the only restrictions on f and d are the ones mentioned above.

## 7. Method of Synthesis

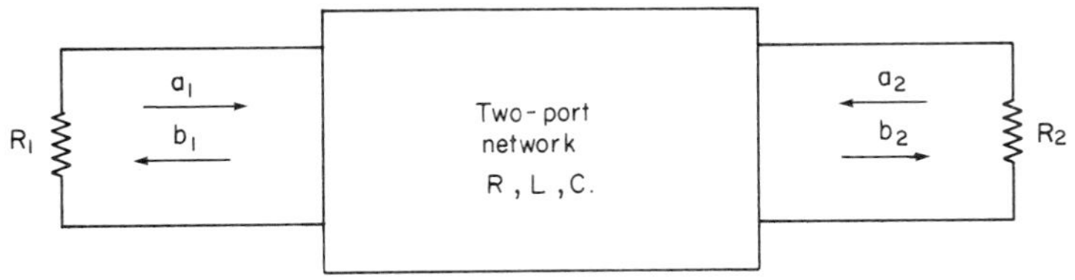


Fig. 5

Normally a network is designed to realise a specified function of the ratio of the power output $P$ to the power output $P_o$ when the source is connected directly to the load. This ratio could be expressed in terms of the scattering parameter $S_{21}$ and the terminal resistors $R_1$ and $R_2$

$$\frac{P_o}{P} = \frac{4R_1R_2}{(R_1+R_2)^2} \frac{1}{|S_{21}(j\omega)|^2} .$$

When $\frac{P_o}{P}$ is given, $S_{21}$ can be determined. Next the input scattering parameter $S_{11}$ is obtained from

$$|S_{11}(j\omega)|^2 = 1 - |S_{21}(j\omega)|^2$$

and

$$S_{11}(s)\, S_{11}(-s) = 1 - |S_{21}(j\omega)|^2_{s=j\omega} .$$

The input impedance can now be found,

$$Z_{in}(s) = \frac{1 + S_{11}(s)}{1 - S_{11}(s)} .$$

Thus we have converted the problem of transfer function synthesis to that of a driving point impedance synthesis. If we restrict our discussion to ladder type networks then the problem is then to develop the function $Z_{in}(s)$ into a continued fraction expansion in terms that can be realised by circuit elements.

## 8. Predistortion

In the method of predistortion the variable $s$ is changed to a new variable $\lambda$ and the desired function $F(s)$ is conformally mapped into the new plane. The function $\tilde{F}(\lambda)$ is then synthesised and the appropriate network obtained. The variable $\lambda$ is chosen in such a way that transforming back to the original frequency variable s one need only modify the circuit elements obtained without any further mathematical steps. This method is useful when it is desired to include losses in the circuit elements. It can also be used when active circuit elements such as negative resistors are to be included. In this case a whole range of amplifier circuits can be synthesised using methods that were only applicable to passive networks.

## 9. Equal Predistortion

This is used when all the circuit elements have the same loss (or gain) factors d. The new frequency variable $\lambda$ is given by

$$\lambda = s + d = \Sigma + j\Omega$$

and the impedance function Z(s) is transformed to

$$\tilde{Z}(\lambda) = Z(\lambda - d) .$$

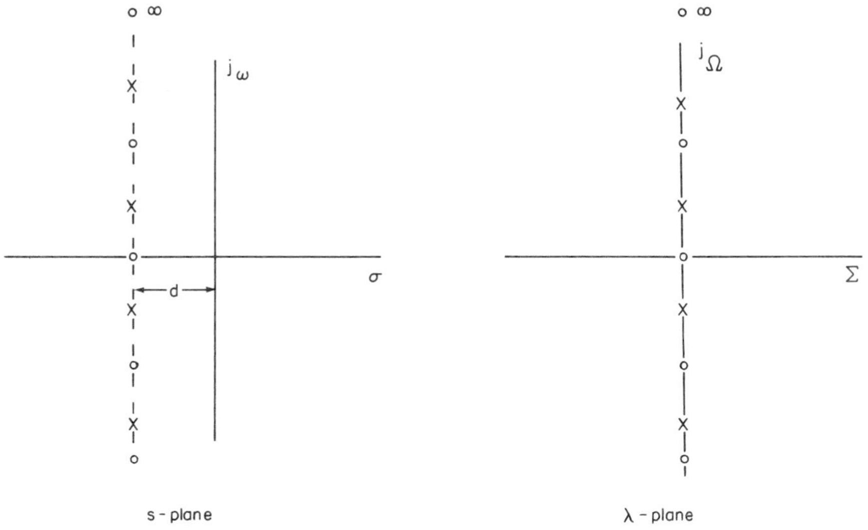


Fig. 6

This transformation is particularly useful in driving point impedance synthesis when all the poles and the zeros lie in a line parallel to the $j\omega$ axis at a distance $-d$ away from it as shown in Fig. 6. The transformed impedance will then represent an L-C network for which a continued fraction expansion can be readily obtained,

$$\tilde{Z}(\lambda) = \frac{1}{\lambda C_1} + \frac{1}{\lambda L_2} + \frac{1}{\lambda C_3} + \dots \frac{1}{\lambda L_n} .$$

To transform back to the original function $Z(s)$ we used the element transformation shown in Table II.

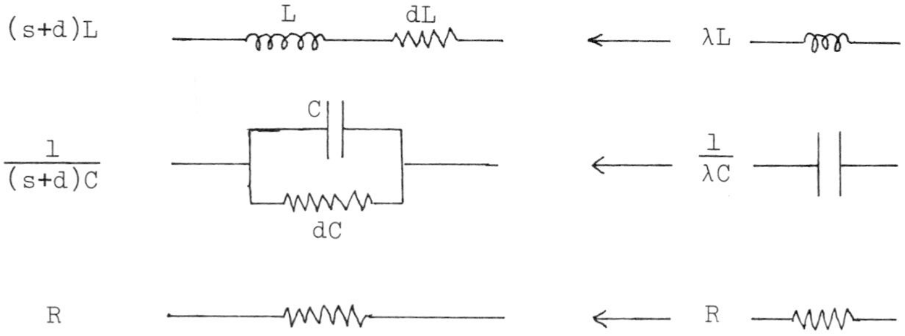


Table II

Other transformations could be used for which the following are examples

$$\lambda = \frac{sd}{d+s} , \quad \lambda = \frac{1}{s+d} , \quad \lambda = \frac{s+d}{sd} .$$

Double and multiple transformations are also possible. It should be noted that the above method allows negative resistors to be used without any change in the synthesis procedure.

The procedure for synthesising transfer functions must be modified when predistortion is used. The steps comparable to those given in Section 7 are listed below.

Given the ratio

$$\frac{P_o}{P} = \frac{4R_1R_2}{(R_1+R_2)^2} \frac{1}{|S_{21}(j\omega)|^2}$$

and changing the variable $s=\lambda+d$ we get

$$\tilde{S}_{21}(\lambda) = S_{21}(\lambda+d)$$

$$S_{11}(j\Omega)^2 = 1 - |S_{21}(j\Omega)|^2$$

$$S_{11}(\lambda)\, S_{11}(-\lambda) = |S_{11}(j\Omega)|^2$$

$$Z_{in}(\lambda) = R_1 \frac{1 + S_{11}(\lambda)}{1 - S_{11}(\lambda)} .$$

$Z_{in}(\lambda)$ is then synthesised and Table II used to obtain the desired network.

## 10. Unequal Predistortion

This transformation allows the loss (or gain) factors associated with all the inductors to be different from those associated with the capacitors. If these are given by $d_L$ and $d_C$ we define $d$ and $\delta$ such that

$$d_L = d + \delta \quad \text{and} \quad d_C = d - \delta .$$

The new variable $\lambda$ is given by $\lambda=s+d$ and the new impedance function $\tilde{Z}(\lambda)$ is given by

$$\tilde{Z}(\lambda) = \sqrt{\frac{\lambda-\delta}{\lambda+\delta}}\; Z(\lambda-d) .$$

The synthesis procedure will then be concentrated on finding a continued fraction expansion for $\tilde{Z}(\lambda)$ in terms of the quantity

$\sqrt{\lambda^2-\delta^2}$ or an expansion of $Z(\lambda-d)$ in alternate terms of $(\lambda-\delta)$ and $(\lambda+\delta)$

$$\tilde{Z}(\lambda) = \frac{1}{C_1\sqrt{\lambda^2-\delta^2}} + \frac{1}{L_2\sqrt{\lambda^2-\delta^2}} + \dots \frac{1}{L_n\sqrt{\lambda^2-\delta^2}}$$

and

$$Z(\lambda-d) = \frac{1}{C_1(\lambda-\delta)} + \frac{1}{L_2(\lambda+d)} + \dots \frac{1}{L_n(\lambda+d)} .$$

Then Table III is used to transform the circuit back to the desired one.

| | | | | |
|---|---|---|---|---|
| $L(s+d_L)$ | 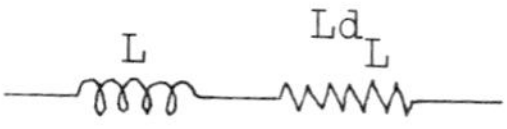  | ← |  | $L\sqrt{\lambda^2-\delta^2}$ |
| $\frac{1}{C(s+d_C)}$ | 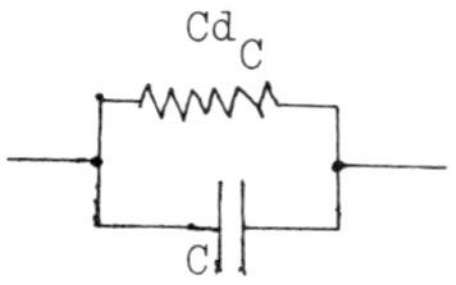  | ← | 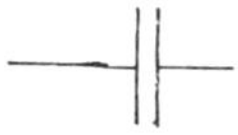 | $\frac{1}{C\sqrt{\lambda^2-\delta^2}}$ |
| $R$ |  | ← |  | $R\sqrt{\frac{\lambda-\delta}{\lambda+\delta}}$ |

Table III

The synthesis procedure of transfer functions is modified for the new transformation and similar steps to the ones described earlier in Section 9 are obtained.

An example of a normalised Butterworth amplifier circuit synthesised using this method is shown in Fig. 7.

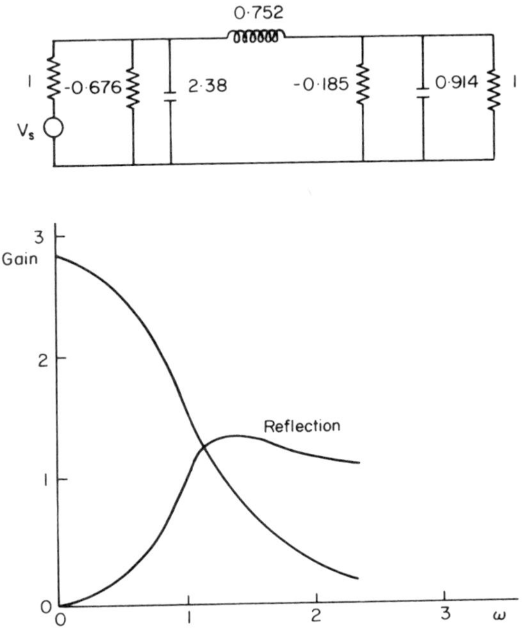


Fig. 7

## 11. Time Domain Synthesis

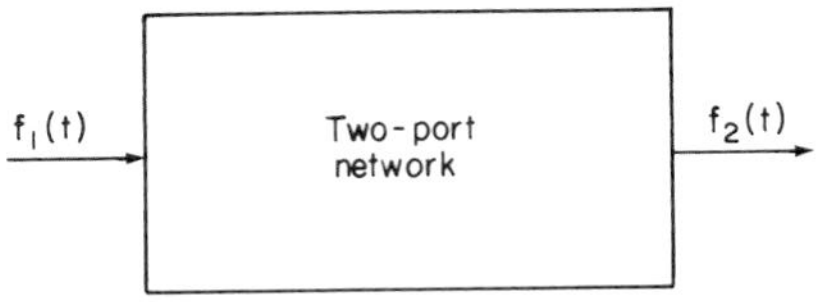


Fig. 8

If $f_1(t)$ is the excitation or input signal and $f_2(t)$ is the response or output signal then the transfer function $H(s)$ in the frequency domain is given by $H(s) = F_2(s)/F_1(s)$ and in the time domain the transfer function $h(t)$ is given by $h(t) = L^{-1}H(s)$. If the input signal $f_1(t)$ is a unit impulse function $\delta(t)$ then $F_1(s)=1$ and $h(t)=f_2(t)$.

In many applications, circuits are designed to have a specified time domain response for a certain input function. Although the time domain and the frequency domain responses are related by the Laplace transform, a frequency domain approximation does not necessarily give the best time domain approximation. In the direct Padé approximant method the frequency domain function $H(s)$ is approximated by the approximant of the appropriate degree $\bar{H}(s)$. This does not give any direct control to the time domain error $\varepsilon(t)$. In the indirect method the approximation is made in relation to a new frequency function $H(z)$ derived from the original function $H(s)$ such that control is obtained over the error $\varepsilon(t)$.

## 12. The Indirect Padé Approximant Method (Teasdale, 1953)

The error in the transfer function approximation in the time domain, $\varepsilon(t)$, is given by

$$\varepsilon(t) = |h(t) - \bar{h}(t)|$$

$$= \frac{1}{2\pi} \left| \int_{-j\infty}^{j\infty} (H(s) - \bar{H}(s))\, e^{st}\, ds \right| .$$

Employing the conformal transformation $s \to z$ where

$$s = \frac{1 - z}{1 + z} ,$$

we transform the right-hand side of the s plane into the inside of a circuit circle with the centre at the origin.

It has been shown (Teasdale, 1953) that with such transformation, the upper bound of the error in the time domain is given by

$$|\varepsilon(t)|^2 \leqslant \frac{2^{2N-3}}{\pi^{3/2}} \frac{\Gamma(N-3/2)}{\Gamma(N-1)} \int_{-\pi}^{\pi} d\theta \left| H_2(z) - \bar{H}_2(z) \right|^2_{z=e^{j\theta}}$$

where

$$H_2(z) = \frac{H_1(z)}{(1+z)^N}$$

$$H_1(z) = H\left(\frac{1-z}{1+z}\right)$$

$$N = \text{number of zeros of } H(s) \text{ at } s=\infty$$

$$\bar{H}_2(z) = \frac{\bar{H}_1(z)}{(1+z)^N} = \text{The approximant of } H_2(z)$$

$$\bar{H}_1(z) = \bar{H}\left(\frac{1-z}{1+z}\right)$$

Thus, $\varepsilon(t)$, the error in the time domain has an upper bound that is independent of time when it is expressed in terms of $H_2(z)$ rather than $H(s)$. To minimise $\varepsilon(t)$ we take $\bar{H}_2(z)$ as the Padé approximant of $H_2(z)$.

The following steps summarise the indirect method

(a) Given $h(t)$ obtain $H(s)$ from $H(s) = h(t)$.

(b) Obtain $H_1(z) = H\left(\frac{1-z}{1+z}\right)$ and $H_2(z) = \frac{H_1(z)}{(1+z)^N}$.

(c) Obtain $\bar{H}_2(z)$, the Padé approximant of $H_2(z)$ satisfying the error upper bound requirement.

(d) Obtain $\bar{H}_2(z)$, $\bar{H}_1(z)$, $\bar{H}(s)$ and $\bar{h}(t)$.

An example has been solved for $h(t) = \frac{J_1(t)}{t}$.

The results of the approximations using the direct and the indirect methods are shown in Fig. 9. In both cases the [1/2] approximant was used.

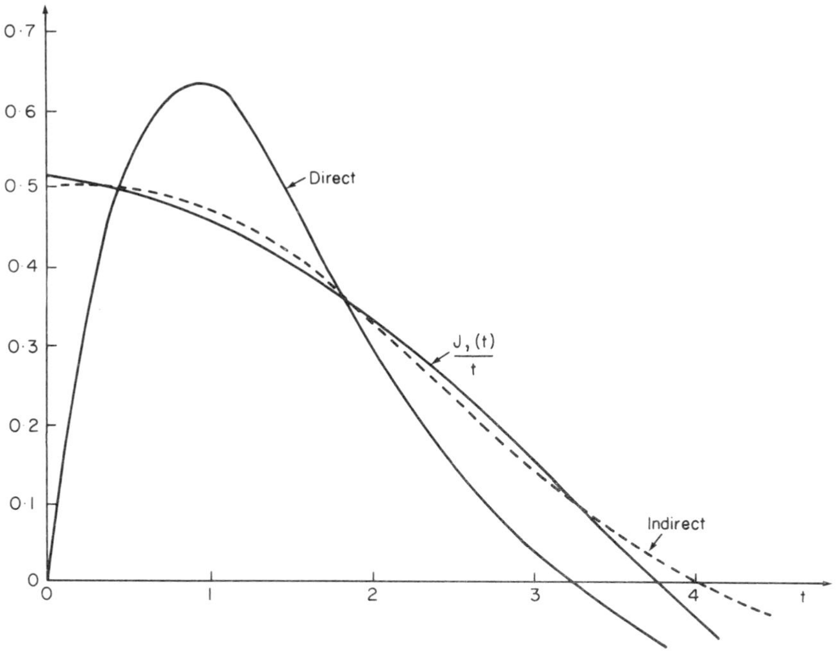


Fig. 9


References

Anderson J. and Lee H. B., (1966). IEEE Trans. on Circuit Theory, Vol. CT-13 No.2, pp.244-258.

Darlington S., (1939). J. Math. Phys., 18, p.257.

Gillemin E. A., (1957). In "Synthesis of Passive Networks", John Wiley, New York.

Newcomb R. W., (1966). In "Linear Multipost Synthesis", McGraw-Hill, New York.

Teasdale R. D., (1953). IRE Convention Record, Vol.1, Pt.5, pp.89-94.

Ulstad M. S., (1968). IEEE Trans. on Circuit Theory, Vol.CT-15, pp.205-210.

Weinberg L., (1962). In "Network Analysis and Synthesis", McGraw-Hill, New York.

# THE SIMULATION OF A CONTINUOUSLY VARIABLE TRANSPORT DELAY

J. B. Knowles, A. B. Keats and D. W. Leggett

(Atomic Energy Establishment, Winfrith, Dorset, England.)

As a result of any form of flow between units, transport delays or distance velocity lags occur in many types of plant. Some of these delays have a marked effect on stability and dynamic performance, so that they must be included in an analogue simulation of the mathematical model of the system. The problem is that of synthesising the transfer function $\exp(-s\tau)$, where the delay $\tau$ is controlled by another simulated variable. After many years experience in the simulation of nuclear plant, the analogue computation group at Winfrith produced the following extracted specification for a continuously variable transport delay unit (Knowles and Leggett, 1972).

| | |
|---|---|
| Step function response time to 1% $\leqslant$ 10 msec<br>Step function response overshoot $\leqslant$ 1% | with any constant delay |
| Delay band 1 - 10 sec over 0 - 1 Hz<br>Delay band 10 - 100 sec over 0 - 0.1 Hz<br>Delay band 0.1 - 1 sec over 0 - 10 Hz | with a steady-state accuracy of 0.1% for any fixed delay in each range. |

Lumped parameter networks involving operational amplifiers, electro-mechanical devices and sampled-data storage schemes are three basic methods of realising such a simulation unit. The performance of a transport delay simulation unit may be assessed in terms of its Delay-Bandwidth product. Clearly, the greater the delay and the higher the operating frequency achieved by the device the better is its design. An unusual aspect of Padé networks is that an arbitrary increase in order fails to reduce the error in the phase-frequency characteristic, when the networks are operated outside a delay-bandwidth product of 6.5. In addition, the step function response of Padé networks has quite unacceptable overshoots.

Magnetic recording systems have been used to generate a transport delay using a servo drive to control the tape velocity between the read/write heads. However, the tape splice generates interference in the form of large voltage spikes and the inertia of the drive motor makes the simulation of rapid changes of fluid velocity impossible.

Such considerations as the above lead naturally to the choice of the completely electronic sampled-data system shown in Fig. 1.

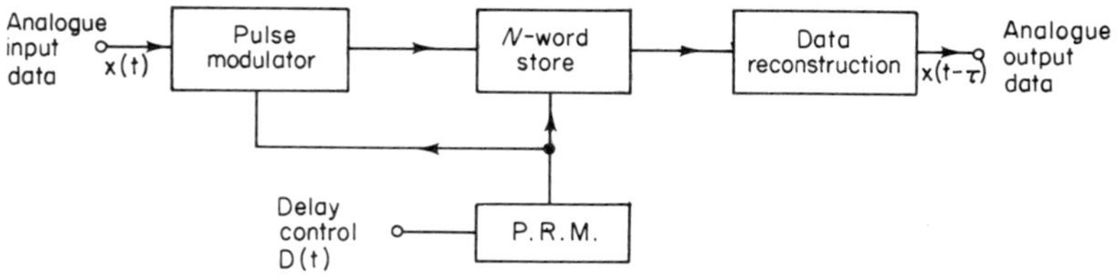


Fig. 1.

The pulse rate modulator produces a pulse train whose repetition note is proportional to the control signal D(t). This pulse train is used to initiate sampling of the input data, and to clock samples through the N-word store. If the signal D(t) changes relatively little over the transition period of a sample through the store, then the delay imposed on the sample is

$$\tau = \frac{\text{Number of Store Words (N)}}{\text{Clock Frequency } (f_s)} \; . \qquad (1)$$

As the delay unit is to be used with an analogue computer, the sampled-data output from the store must be reconstructed into analogue format. Hence, the sampling frequency ($f_s$) must be sufficiently high for this reconstruction to be performed to the accuracy demanded by the specification. An analysis shows that this accuracy can be achieved much more economically by the use of linear interpolation than by simple clamping of the delayed samples. Further cost reductions are achieved by the use of MOS shift registers in preference to magnetic core or analogue storage elements.

Used as a constant delay element, the delay bandwidth product of this delay unit is about a decade higher than the asymptotic maximum value (6.5) for Padé networks. Apart from simulating $\exp[-s\tau(t)]$, the system can be used as a programmable function generator. Logic has been developed (Keats and Leggett, 1972) which enables analogue voltages or punched tape data to be encoded into the store. By recirculating this data around the store, the analogue function can be reproduced at will almost indefinitely.

References

Keats A. B. and Leggett D. W., (1972). The Radio and Electronic Engineer 42, 179.

Knowles J. B. and Leggett D. W., (1972). The Radio and Electronic Engineer 42, 172.

APPROXIMATION OF LINEAR TIME-INVARIANT SYSTEMS.

Y. Shamash

(Department of Computing and Control, Imperial College, London)

## 1. Introduction

In the study of control systems it is often desirable to approximate a high order, linear, time-invariant system by a system of lower order.

Two methods of reduction, widely used in engineering circles, are the method of Chen and Shieh (1968), and the method of equating moments (Lees, F.P. and Gibilaro, L.G., 1969), which are actually equivalent and a special case of Padé approximation (Shamash, Y. 1972).

In this paper a new method of reduction is introduced which retains the dominant modes of the system in the reduced model. The method is based on the concept of Padé approximation about two (or more) points, first introduced by Baker, et. al (1961). Koenig's generalized theorem (Householder, A.S., 1970) for the convergence of poles of Padé approximants is of great value in this method since it can be used to compute the number of dominant modes of the system and their location.

## 2. Koenig's Generalized Theorem

The theorem is stated here. The proof may be found in Householder, 1970.

Theorem

Let $f(s) = C_o + C_1 s + C_2 s^2 + \ldots, \; C_o \neq o$ (2.1)

be meromorphic for $|s| < R$, and let it have exactly p poles $r_1, r_1, \ldots, r_p$, not necessarily distinct, in this disc.

Let $o < |r_1| < |r_2| < \ldots < |r_p| < \sigma R < R$, and let

$$\Psi(s) = (1-r_1^{-1}s)(1-r_2^{-1}s) \ldots (1-r_p^{-1}s)$$
$$= 1 + a_1 s + a_2 s^2 + \ldots + a_p s^p$$

Finally, let the denominator of the (p,v) Padé approximant be

$$K_v(s) = 1 + \alpha_1^{(v)} s + \alpha_2^{(v)} s^2 + \ldots + \alpha_p^{(v)} s^p \quad (2.3)$$

Then $\alpha_i(v) = a_i + O(\sigma^v)$ and (2.4)

$$K_v(s) = \Psi(s) + O(\sigma^v). \quad (2.5)$$

## 3. Padé approximation and dominant mode reduction

The main feature of most simplification procedures is the suppression of certain modes associated with the larger eigenvalues of the high order system. Consider the following high order transfer function.

$$G(s) = \frac{d_o + d_1 s + d_2 s^2 + \ldots + d_{n-1} s^{n-1}}{(s+s_1)(s+s_2) \ldots (s+s_n)} \quad (3.1)$$

$$= C_o + C_1 s + C_2 s^2 + \ldots \quad (3.2)$$

Let the reduced model be of the form

$$R(s) = \frac{a_o + a_1 s + \ldots + a_{k-1} s^{k-1}}{b_o + b_1 s + \ldots + b_{k-1} s^{k-1} + s^k} \quad (3.3)$$

Then for R(s) to be a Padé approximant of G(s), we have

$$a_o = b_o C_o$$

$$a_1 = b_o C_1 + b_1 C_o \qquad (3.4)$$

$$\cdot \quad \cdot \quad \cdot \quad \cdot \quad \cdot \quad \cdot$$

$$0 = b_o C_{2k-1} + b_1 C_{2k-2} + \ldots + C_k$$

Now if R(s) is to retain the pole at $s=-s_1$ (say), then the last equation in (3.4) is replaced, using Padé approximation about two points, by

$$0 = b_o - b_1 s_1 + b_2 s_1^2 + \ldots + (-1)^k s_1^k$$

Hence these equations can then be solved for the coefficients $b_i$, $a_i$ $(i = 0,1,2,\ldots,k-1)$ of equation (3.3).

More generally, if R(s) is required to retain the k dominant poles (poles nearest the origin) of the high order system, then the $b_i$'s are uniquely determined by the location of these poles and the $a_i$'s $(i=0,1,\ldots,K-1)$ are determined using the first K equations of (3.4).

So far it has been assumed that the dominant poles of the system are known, which in most cases is not necessarily true. This is where Koenig's theorem is of great use. For, suppose that a high order system transfer function has the pole locations shown in Fig. 1.

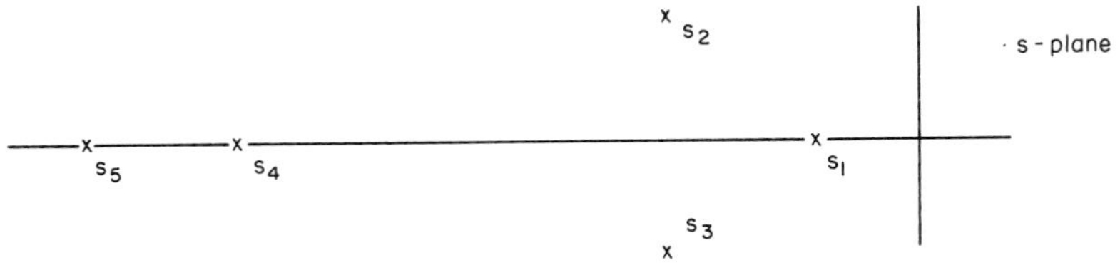


Fig. 1. Location of Poles of G(s)

Then applying Koenig's theorem to the expansion (3.2), the first pole nearest the origin, $s_1$, can be picked out. If $s_1$ is close to $s_2$ and $s_3$ then (2.4) and (2.5) will converge very slowly, implying that there is at least one other pole close to $s_1$. The theorem is then applied to test for the convergence of two poles. In this example it will not converge since $s_2$ and $s_3$ are complex conjugate. Hence the theorem is applied to test for the convergence of the first three poles. If the convergence is very fast, then it will imply that the system has three dominant poles and the equations which they satisfy can be easily computed (Householder, A.S., 1970).

## 4. Conclusion

A new method of reduction has been introduced. The high order transfer function is first expanded into a power series, about s=o. Then Koenig's theorem is used to determine the number of dominant poles and their location. Finally Padé approximation is used to determine the numerator coefficients of the reduced order transfer function. This method has the following advantages:

1. It is very easy to use.
2. Computationally it is easier than other methods which retain the dominant modes.
3. Account is taken of the neglected poles, unlike other methods which simply neglect them.

4. The reduced model is always stable.
5. It is very easily extended to the problem of reducing a matrix transfer function.
6. Good approximation is obtained even if the system doesn't have a set of dominant poles.

A more detailed report containing examples is available from the author.

Acknowledgements

The author is grateful to the University of London for a Studentship.

## 5. References

Baker, G.A. Jr., Gammel, J.L. and Wills, J.G., (1961), J Math. Analysis and Applications, Vol. 2, pp. 405-418.

Chen, C.F. and Shieh, L.S., (1968), JACC Record, pp 454-461.

Householder, A.S., The numerical treatment of a single non-linear equation, McGraw-Hill Book Company, New York, 1970.

Lees, P.F. and Gibilaro, L.G., (1969), Chem. Eng. Science, Vol. 24, pp. 85-93.

Shamash, Y., (1972), Department of Computing and Control Publication 72/6 (Research Report), Imperial College of Science and Technology, London SW7 2BT.

# FREQUENCY DOMAIN APPROXIMATION TECHNIQUE FOR OPTIMAL CONTROL

S. C. Chuang

(Department of Computing and Control, Imperial College, London)

## 1. Introduction

Extensive study has been made of the problem of optimising a linear stationary control system with respect to a quadratic cost function in the frequency domain. The solution requires the spectral factorisation of a positive definite matrix often known as spectral density matrix (SDM) which is often very tedious, particularly for multivariable control systems (D. C. Youla, 1961, M. C. Davies, 1963 and B. D. O. Anderson, 1967).

We shall demonstrate, in this paper, that an approximation technique of the Padé type can be applied to the SDM P to obtain a low-order approximation so that the corresponding suboptimal control to the system can be obtained easily. An iterative algorithm, in the frequency domain, will also be presented. Several useful features of the algorithm will be described. A numerical example will be presented.

## 2. Problem Formulation

Central to the solution of the above optimisation problem is the problem of spectral factorisation, i.e. finding a matrix G(s) such that

$$G^T(-s)\ G(s) = R + W^T(-s)\ QW(s)$$

where W(s) is the transfer function matrix of the system; $R>0$ and $Q\geqslant 0$ are the weighting matrices of the quadratic cost function.

## 3. Suboptimal Control

In order to obtain good approximations during the initial transient and at the steady-state, Padé approximation about s=0 and s=∞ is applied to the SDM P; i.e. if $\bar{P}$ is a low-order approximation to P, then in terms of power series, we have

$$P - \bar{P} = A_n s^n + A_{n+1} s^{n+1} + \ldots \qquad \text{about } s=0$$

$$= B_{n'} s^{-n'} + B_{n'+1} s^{-(n'+1)} + \ldots \qquad \text{about } s=\infty ,$$

for some positive integers n and n'.

## 4. Iterative Algorithm

We shall now develop an iterative algorithm which will prove to be particularly useful for multivariable systems. Suppose that the $i^{th}$ approximation to the SDM P is $P_i$ where $P_i(s) = G_i^T(-s)G_i(s)$ and $G_i^{-1}(s)$ exist. Then,

$$P = P_i + (P - P_i)$$

where $P-P_i$ is the error incurred by using the approximation $P_i$ instead of P. It is obvious that

$$P = P_i + G_i^T(-s)V_iG_i(s) \qquad (1)$$

where

$$V_i = (G_i^T(-s))^{-1}P\,(G_i(s))^{-1} - I\ . \qquad (2)$$

By using Padé approximation about s=0 and s=∞, a low-order approximation $\bar{V}_i$ to the rational matrix $V_i$ can be found and

$$P = P_i + G_i^T(-s)\,\bar{V}_iG_i(s) + G_i^T(-s)(V_i-\bar{V}_i)\,G_i(s).$$

By writing $P_{i+1} = P_i + G_i^T(-s)\,\bar{V}_iG_i(s)$, we obtain a new approximation $P_{i+1}$ to P. Intuitively, $P_{i+1}$ would be a better approximation to P than $P_i$, as $P_{i+1}$ includes an additional term approximating the error $P-P_i$. Also, as

$$P_{i+1} = P_i + G_i^T(-s)\,\bar{V}_i G_i(s) = G_i^T(-s)(I+\bar{V}_i)\,G_i(s) ,$$

the spectral factorisation of $P_{i+1}$ requires only that of the low-order matrix $I + \bar{V}_i$.

Assuming that $P_i$ is an approximation such that

$$P - P_i = \sum_{i=n}^{\infty} \gamma_i\, s^i = \sum_{i=n}^{\infty} \theta_i\, s^{-i}$$

for some $n,n'>0$ and expressing $G_i(s)$, $G_i^{-1}(s)$ in terms of power series, it is easy to see that

$$V_i = \sum_{i=n}^{\infty} \xi_i\, s^i = \sum_{i=n'}^{\infty} \delta_i\, s^{-i} .$$

It follows that the difference between $V_i$ and its approximation $\bar{V}_i$ is

$$V_i - \bar{V}_i = \sum_{i=n+m}^{\infty} \xi_i'\, s^i = \sum_{i=n'+m'}^{\infty} \delta_i'\, s^{-i}$$

for some $m,m'>0$, and hence

$$P - P_{i+1} = \sum_{i=n+m}^{\infty} \gamma_i'\, s^i = \sum_{i=n'+m'}^{\infty} \theta_i'\, s^{-i} .$$

We have thus shown that, when expressed in terms of power series about $s=0$ ($s=\infty$), $P_{i+1}$ generated by the algorithm agree with P up to a higher order term in $s(s^{-1})$ than $P_i$.

## 5. Example

Consider a scalar 10th order system having the transfer function W(s) where

$$W(s) = \frac{a(s)}{b(s)} .$$

$$a(s) = .5057\times10s^9 + .4697\times10^3s^8 + .15707\times10^5s^7 + .26459\times10^6s^6$$
$$+ .25547\times10^7s^5 + .14913\times10^8s^4 + .53105\times10^8s^3 + .1116\times10^9s^2$$
$$+ .1258\times10^9s + .576\times10^8$$

$$b(s) = s^{10} + .109\times10^3s^9 + .4363\times10^4s^8 + .88875\times10^5s^7 + .1051404\times10^7s^6$$
$$+ .7674576\times10^7s^5 + .25302112\times10^8s^4 + .10134752\times10^9s^3$$
$$+ .1732352\times10^9s^2 + .158112\times10^9s + .576\times10^8;$$

Q=10, R=1; initial conditions $y_o(s)=0$, desired system response $y_d(s) = \frac{1}{s}$. The SDM P is then a rational regular function of 20th order. It is then approximated by a 2nd order rational regular function, using Padé approximation about s=0 and s=∞. Comparison between the optimal control (curve a) and the suboptimal control (curve b) is shown in Fig. 1.

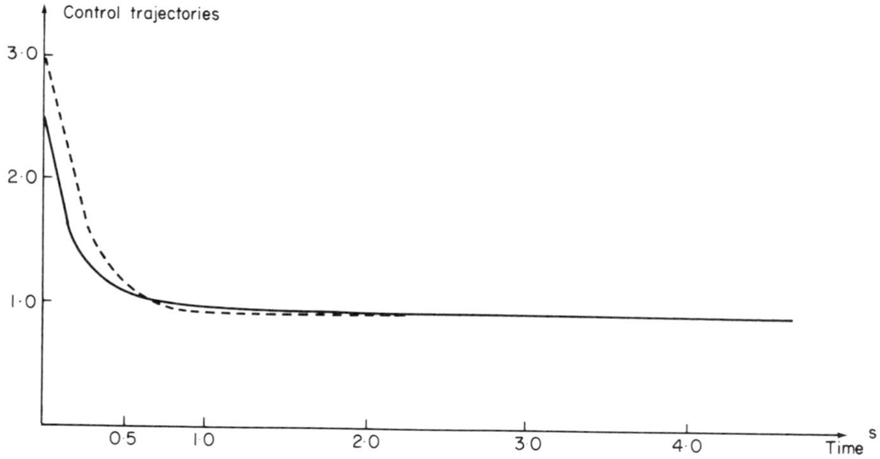


## References

Anderson B. D. O., (1967). IEEE Trans. AC-12, 410.
Davies M. C., (1963). IEEE Trans. AC-8, 296.
Youla D. C., (1961). IEEE Trans. IT-7, 172.

# A PADE CHEBYSHEV APPROXIMATION IN NETWORK THEORY

H. P. Debart

(C. G. E., Marcoussis, France)

## 1. Introduction

In network theory, the problem of approximating a given function of the frequency by a rational fraction is sometimes approached by a Padé method. However, a major problem is often to derive a Chebyshev approximation of a given transfer function on a given interval of the frequency axis.

The approach we are developing consists of a mapping of the frequency axis onto the unit circle, and of achieving on this circle a Chebyshev approximation; this last step is possible if the given function is replaced by a Padé approximant.

## 2. Conformal mapping on the unit circle and Chebyshev approximation

If the approximation area is the whole axis, as in the most elementary case, the transformation $p = \frac{1-z}{1+z}$ is used, where $p = i\omega$ is the complex frequency. We will apply the following theorem :

Let $z_1 \ldots z_n$ be complex numbers whose moduli are greater than 1. Consider the set C of polynomials of degree n, in which the coefficient of $z^r$ is $A_{n-r}$ and the coefficient of $z^n$ is unity. Consider the quantity:

$$\left| \frac{z^n + A_1 z^{n-1} + \ldots + A_n}{(z-z_1)\ldots(z-z_n)} \right|$$

As the complex variable z goes around the unit circle, this modulus reaches somewhere a maximal value. There is one (and only one) polynomial that makes this maximal value minimal. It has the form

$$z^n + A_1 z^{n-1} + \dots + A_n = (z - \frac{1}{z_1^*}) \dots (z - \frac{1}{z_n^*}) \equiv P(z)$$

and the corresponding minimum is

$$\text{Min}_C \ \text{Max} \left| \frac{z^n + A_1 z^{n-1} + \dots + A_n}{(z-z_1) \dots (z-z_n)} \right| = \frac{1}{|z_1 z_2 \dots z_n|}$$

If the variable z goes around the unit circle, the modulus of each ratio

$$\frac{z - \frac{1}{z_i^*}}{z - z_i}$$

is constant, equal to $\frac{1}{|z_i|}$. Hence the modulus of the product of these ratios is constant and remains equal to its maximal value. On the other hand, the argument of the product of these ratios increases by $2\,n\pi$, as the variable z goes around the unit circle. Hence the real and imaginary parts oscillate between $\pm \frac{1}{|z_1 \dots z_n|}$, and have interwoven zeroes.

The number of zeroes for both of them is 2n.

## 3. Padé Chebyshev approximation

Let f(p) be a transfer function of a physical, minimal phase-shift network. Let f(p) be approximated by a rational function of p, $\frac{n(p)}{d(p)}$, with criterion that $1 - f(p)\,\frac{d(p)}{n(p)}$ approximate 0 on the frequency axis, in the Chebyshev sense (this is the fundamental problem in filter design, for example).

The conformal mapping $p = \frac{1-z}{1+z}$ will change f(p) to F(z) and $\frac{d(p)}{n(p)}$ to $\frac{D}{N}(z)$. We replace the function F(z) by a Padé approximant: $F(z) \sim \frac{N_0}{D_0}(z)$ (N and D are polynomials of degree m, $N_0$ and $D_0$ are polynomials of degree n, and n >> m. For example F(0) = 1.

$$N_0 = 1 + a_1 z + \dots + a_n z^n \qquad N = C_0 + C_1 z \dots + C_n z^m$$

$$D_0 = 1 + b_1 z + \dots + b_n z^n \qquad D = 1 + d_1 z \dots + d_m z^m$$

The solution of the problem has been given above: the roots of the polynomials $D_0N - N_0D$ and $D_0N$ are to be reciprocals of each other. This condition is converted into m + n relations between the 2(m+n)+1 coefficients. On the other hand, the fraction $N_0/D_0$ has to approximate the given function in the Padé sense. The Padé identity supplies the (m + n + 1) further relations we need to compute the coefficients.

## 4. Numerical treatment of an example

Let us choose $f(p) = \exp\left(\frac{1-p}{4+6p}\right)$, hence $F(z) = \exp\left(\frac{0.2z}{1-0.2z}\right)$. The [2,2] Padé approximant to F(z) is $\frac{1-2.7z + .263z^2}{1-2.9z + .783z^2}$

The Chebyshev approximant is the function $\frac{1 - .447z}{.912 - 0.357z}$

The value of the modulus $\left|1 - \frac{N_0}{D_0} \cdot \frac{D}{N}\right|$ which defines the accuracy of the approximation is: 0.255.

## 5. General resolution of the equations

In achieving any approximation of this type, 2m + 2n + 1 equations are to be solved. Among them m + n + 1 are linear and m + n are of the following form :

$$\frac{a_0 x_1 + \ldots + a_N x_N}{b_0 x_1 + \ldots + b_N x_N} = \frac{a'_0 x_1 + \ldots + a'_N x_N}{b'_0 x_1 + \ldots + b'_N x_N} =$$

$$\ldots = \frac{a^{(g)}_0 x_1 + \ldots + a^{(g)}_N x_N}{b^{(g)}_0 x_1 + \ldots + b^{(g)}_N x_N}$$

If we let $\lambda$ be the common value of these ratios, we get $m + n + 2$ linear equations, with a condition of compatibility in $\lambda$. Corresponding equation gives eigenvalues and from each of them an approximation can be derived. The most favourable one is to be chosen, which is the approximation that minimises

$$\left| 1 - \frac{N_0}{D_0} \cdot \frac{D}{N} \right|$$